Glamorous Residence

新魅样板

深圳市创扬文化传播有限公司 编

赵 欣 译

大连理工大学出版社

Dalian University of Technology Press

图书在版编目(CIP)数据

新魅样板：汉英对照 / 深圳市创扬文化传播有限公
司编；赵欣译. — 大连：大连理工大学出版社，
2011.6

ISBN 978-7-5611-6201-9

Ⅰ. ①新… Ⅱ. ①深… ②赵… Ⅲ. ①住宅—室内装
饰设计—作品集—中国—现代 Ⅳ. ①TU241

中国版本图书馆CIP数据核字（2011）第078803号

出版发行：大连理工大学出版社
　　　　　（地址：大连市软件园路80号　邮编：116023）
印　　　刷：利丰雅高印刷（深圳）有限公司
幅面尺寸：230mm×300mm
印　　张：31
插　　页：4
出版时间：2011年6月第1版
印刷时间：2011年6月第1次印刷
策　　划：袁　斌
责任编辑：刘　蓉
责任校对：李　楠
封面设计：连　帅

ISBN 978-7-5611-6201-9
定　　价：348.00元

电　话：0411-84708842
传　真：0411-84701466
邮　购：0411-84703636
E-mail：designbooks_dutp@yahoo.cn
URL：http://www.dutp.cn

如有质量问题请联系出版中心：（0411）84709246　84709043

Glamorous Residence

GLAMOROUS RESIDENCE

新魅样板

CONTENTS

目录

Hui Jiang Holiday Garden Type A, Nanguo Peach Garden, the South China Sea

南海南国桃园汇江假日花园A型

项目地点：佛山
项目面积：约100平方米
设计师：王启贤
设计公司：王启贤设计事务所（香港）
主要材料：镜钢、木纹石、木面

This project brings a fresh decoration concept, which features the application of bright and clean mirror steel materials that have strong reflexivity, so that the sense of space is enhanced, and at the same time a tinge of mystery is added. In addition to that, a large amount of gray and black are employed, and in the meantime, bright colors are embellished as a set-off. The match of black, white and gray with bright colors is an everlasting popular element, which is highly personalized. Apart from that, the carpet and the bedroom, however, choose the cozy warm-toned red colors, and the European styled classical patterns, such as the spiral lamination and arc patterns, are mix-and-matched in the furniture and furnishings that are of modernity; such a style gives emphasis to the harmonious application of proportion and colors. This project holds a good treatment for details with simplified and smooth lines, which fully reveals the fashionable and modern air.

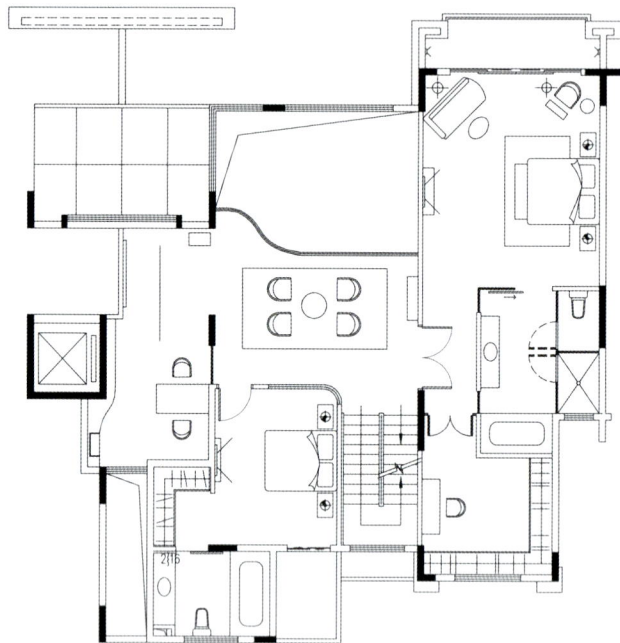

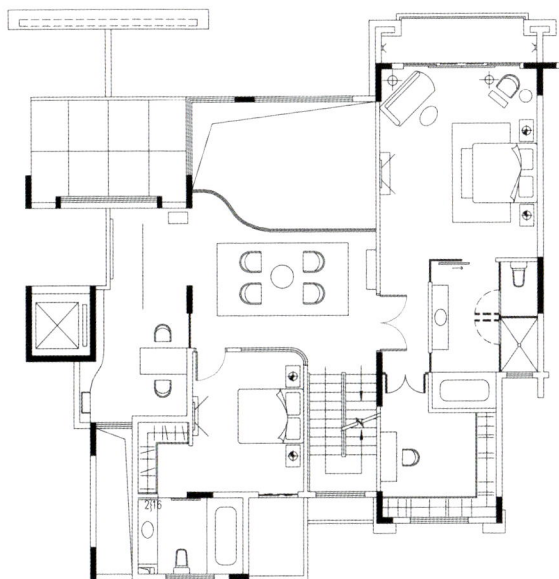

本设计带来了一种新的装饰理念。主要特点是采用光洁的反射度较强的镜钢材料，这样可以使空间感更加强烈，同时又带有一丝的神秘感。另外，设计大量采用灰色、黑色，同时点缀亮色做衬托。黑白灰和亮色的配搭是永恒的流行元素，个性十足，而地毯与卧室又采用温馨的红色暖色调，螺旋纹、弧线等欧式古典纹饰混搭在具有现代感的家具和陈设中，这样的风格强调比例和色彩的和谐运用。设计对细节的把握十分到位，线条简约流畅，尽显时尚、现代气息。

Jindu Golf Art Villa, Hangzhou

杭州金都高尔夫艺墅

项目地点：杭州
项目面积：277平方米
设计师：刘伟婷
设计公司：刘伟婷设计师有限公司

At the entrance, the high dome shows out the owner's transcendent status. Through the carving and polishing of details, the designer fully shows the strong British flavor and classical court style in the living room. The exquisite handmade furniture is British Victorian architectural style, and the huge baroque crystal curtain wall is decorated with delicate carvings, attracting your eyes. The master bedroom creates a romantic style with noble flavor, and the complex decoration, carvings, stable headboard with beautiful lines, and arch-type inlaid carvings reveal the owner's identity and taste. The banquet on the first floor underground is gorgeous and elegant, and the carpet of Indian flavor is an important decoration. The second floor underground plays many function spaces such as audio-visual room, yoga room, and others.

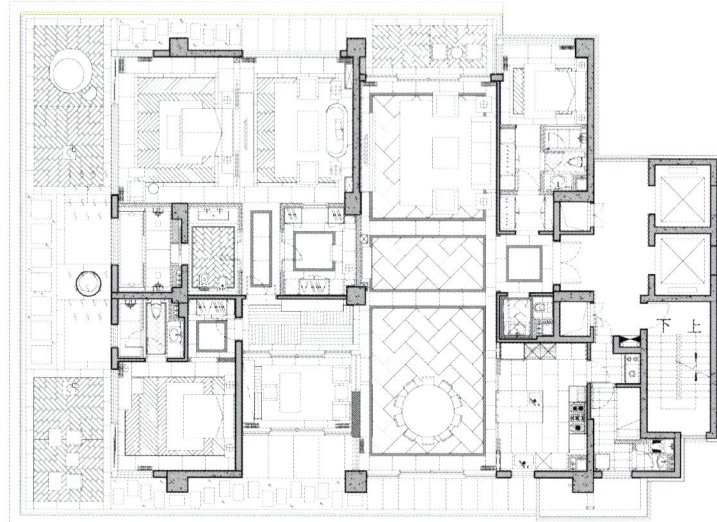

　　玄关处高高的圆顶显现出屋主的超然地位。设计师通过对细节的雕琢，将客厅浓浓的英伦风情及古典宫廷气派进行了充分的展现。精致的手工家具采用了英国维多利亚式建筑风格。客厅中巨幅巴洛克式水晶幕墙面的雕花细腻精美，把客人的目光汇聚于此。主卧室营造了一份浪漫且极具贵族风范的格调，繁复的装饰、雕花、线条优美且稳重的床头板、圆拱形镶嵌着图案的雕刻品，彰显着主人的身份及品位。地下一层宴会厅华丽典雅，充满印度风情的地毯是重要的装饰品。地下二层汇集了影音室、瑜伽运动室等诸多功能空间。

Phoenix City in the World

天地凤凰城

项目地点：廊坊
项目面积：400平方米
设计师：张赫
设计公司：尚层装饰（北京）有限公司
主要材料：实木护墙板、大理石、玻璃、乳胶漆

The design style of the project is mainly the pure style of modern city with the method of combining the visional and the actual, showing a simple, crisp and generous living environment. The design reinterprets the new life of city, combining the fashion and convenience.

In the living room, the three-side sofa surrounds a glass tea table, making you feel relaxed and friendly, and the natural light from the large windows forms an effect of lighting and shadow, enhancing the dynamic space. The window screen with cotton and linen texture meets the requirement of permeability of the space and appropriately overshadows the excessively strong light. The dining room is open design, and the raised floor brings a sense of level into the whole space. The space on the second floor is simple and neat, and the sufficient light and unobstructed line of sight in the bedroom mabe you can enjoy the panoramic view of the scenery outside the window.

本案设计风格以纯净的现代都市风格为主，运用虚实结合的方法，体现简洁、明快、大气的居住环境。设计重新诠释了都市新生活，将时尚与便利结合。

客厅三面环绕的沙发围绕着玻璃茶几，让人感到放松与亲切，客厅利用大玻璃窗的自然光形成的光影效果增加了空间的动感。棉麻质感的纱窗既满足了空间通透性的要求，又适当地遮蔽了过强的光线。餐厅采用开放式设计，筑高的地面为整体空间带来层次感。二楼空间整洁利落，卧室内充足的光线及无遮挡的视线可将窗外的风景尽收眼底。

Tianhua Meidi Show Flat 1008

天华美地样板房1008

项目面积：190平方米
设计师：彭东生
设计公司：汕头天顺祥设计工作室
主要材料：柚木板、墙纸、灰木纹石
摄影师：邱小雄

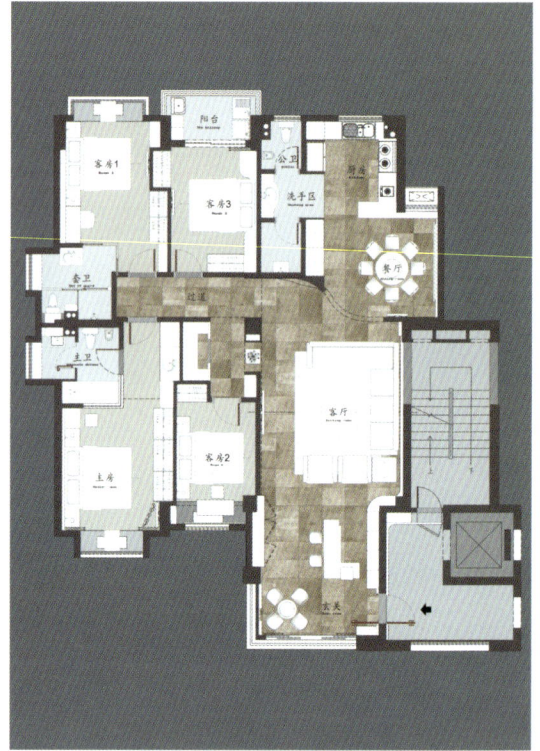

In the design, the designer's first consideration is the need for function, so the storage function is maximized in the layout of the show flat.

The static and dynamic areas are clearly divided in the space, and the layout of the living room is simple and neat without traditional sofa background wall or TV back ground wall, but the copper bell hanging on the pillar, and the branches and some copper birds are slanted between the wall and the pillar to make the space amusing. The open-style kitchen is connected with the dining room into a single entity, and the small platform between them plays a role of dividing. The basin in the public washroom is open style, which is convenient to use, making the whole space more spacious and blurring the boundaries between functional areas. The whole space is divided by lighting effect with clearer gradation.

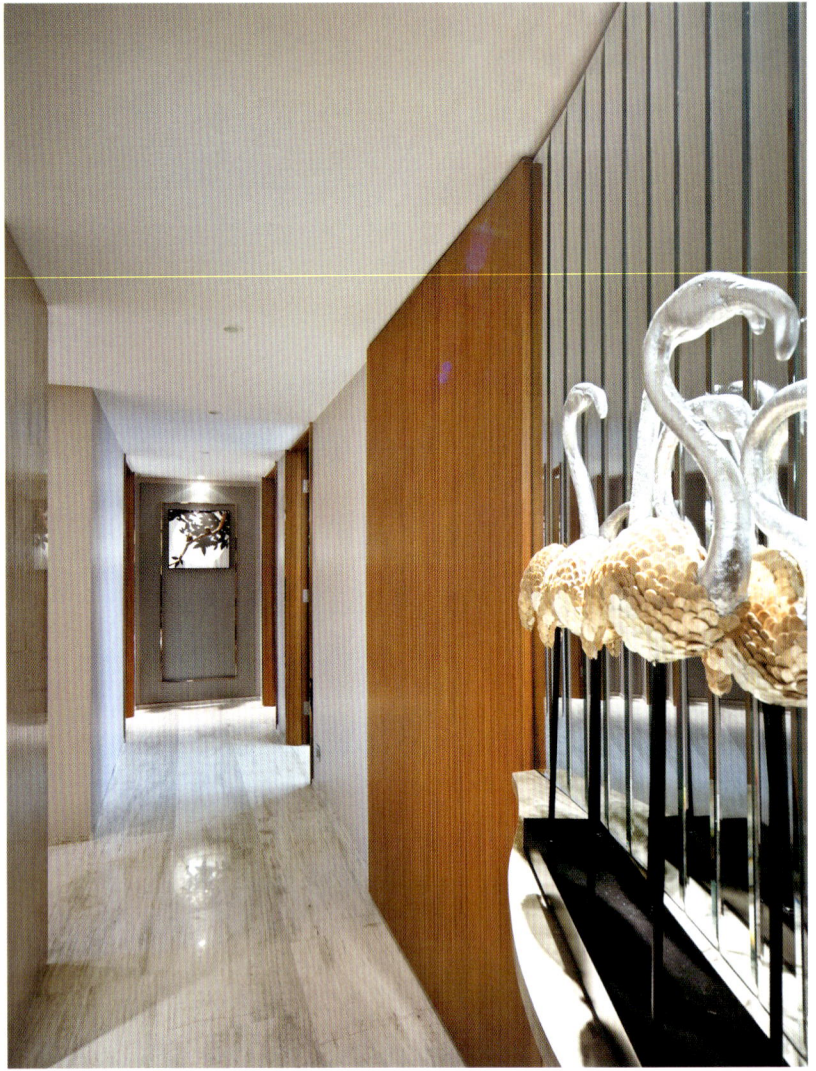

设计师在设计中首先考虑的是功能的需要，使这套样板房在布置上达到了储藏功能的最大化。

室内的动静区区分比较明确。会客厅布置简洁，没有传统的沙发背景墙、电视背景墙，柱子上挂的小铜钟、墙和柱子之间斜挂的树枝和几只铜鸟，都使空间趣味横生。开放式的厨房和餐厅连为一体，中间小吧台起到了间隔的作用。公共卫生间的洗手盆处采用了敞开式处理，既方便了使用，又使整个空间更加宽敞，模糊了功能区的界线。整个区域通过灯光效果进行分隔，使层次更加分明。

Tianhe Completed Design A10

天荷A10实品屋

项目地点：台中
项目面积：165平方米
设计师：杨焕生
参与设计师：王莉莉、王慧静
设计公司：杨焕生建筑室内设计事务所
主要材料：木材、石材、烤漆、进口马赛克、裱布、订制家具
摄影师：刘俊杰

The project introduces low-key colors on the overall design, and the black, white and gray colors are collocated and used to the right, so that the space is neither ostentatious nor heavy. The design focuses on the grasp of the overall qualified style, and the strong neutral colors fill into each room, coupled with dark purple materials and flower patterns in different details, adding a few luxurious flavors into the space. The materials full of texture are used to highlight the quality of the space, and the TV background wall is decorated with stone material of natural cracks, generous without losing modern style. At the same time, the design focuses on the craftsmanship on the details, in the living room, the bird-cage-form decoration enriches the character of the space, so the lines of the space are full of changes.

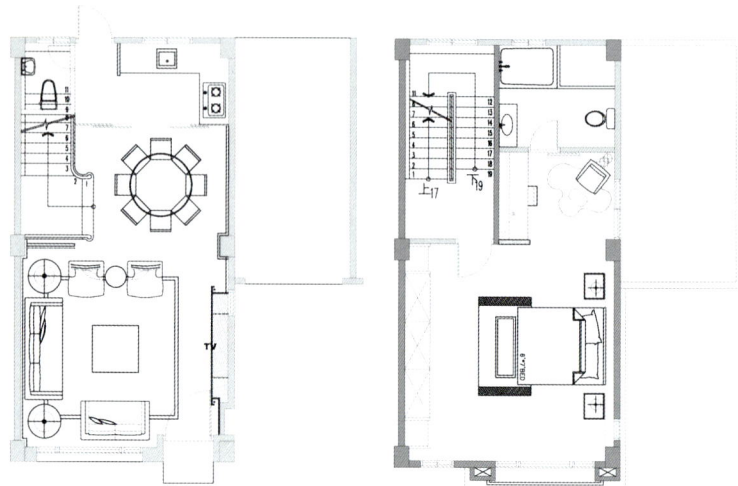

本案整体采用了低调的色彩，设计师将黑白灰三种基础色彩搭配使用得恰到好处，使空间既不浮华也不沉重。设计注重对整体品质风格的把握，浓重的中性色彩充斥了每个房间，在不同的细部搭配上，暗紫色的材质与花卉的图案又使室内平添了几分奢华的情调。设计采用了极富质感的材料，彰显了居室的品质。电视背景墙以天然裂纹形式的石材进行装饰，大气而不失现代风范。设计同时注重对细节的雕琢，客厅鸟笼形式的装饰丰富了空间的性格，更使空间的线条充满了变化。

Old Peak Road

旧山顶道

项目地点：香港
设计师：邓子豪、叶绍雄
设计公司：天豪设计有限公司

Due to job specification, the owner doesn't need plenty of storage space as he needs to go on business trips frequently. Therefore, this apartment is simply kept as a temporary resting place for the owner, so there is much more flexibility on designing this project. The designer uses natural green and wood colors to match the green mountain scenery outside the window. In addition, there are many different "tree" elements. The designers remove the wall between living room and study room, replaced by a glass with tree graphic. Under the lamp light, the shadows of branches are cast on the floor. The bookcase comprises with white and green, together with the green boxes in living room, so that this apartment is full of colors and layers.

The design of the room is rather modest, where the white wardrobe with tree graphics dissected by computer matches well with the side table in wood flooring color, so that the bedroom is full of the cozy and comfortable atmosphere.

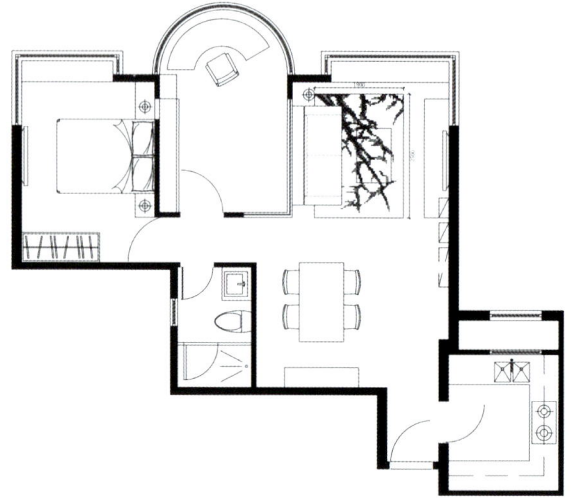

由于业主的工作性质需要经常出差，此住宅只作为业主在香港时的暂休处，因此室内并不需要很多的储物空间，这样便增大了设计上的弹性。

设计师利用了大自然的绿色及木色，以配合窗外的翠绿山景。此外，更渗入了不同的"树"元素。设计师把书房与客厅的墙打通，换上具有树枝图案的玻璃。在灯光下，树枝的影子可以投射到地面上。书房的书柜使用白色和绿色打造，呼应客厅的绿色储物柜，使整个空间都富有色彩和层次。

房间的设计相对平实，白色衣柜以用计算机切割的树图案搭配跟木地板配色的床头柜，使卧室充满温馨的氛围。

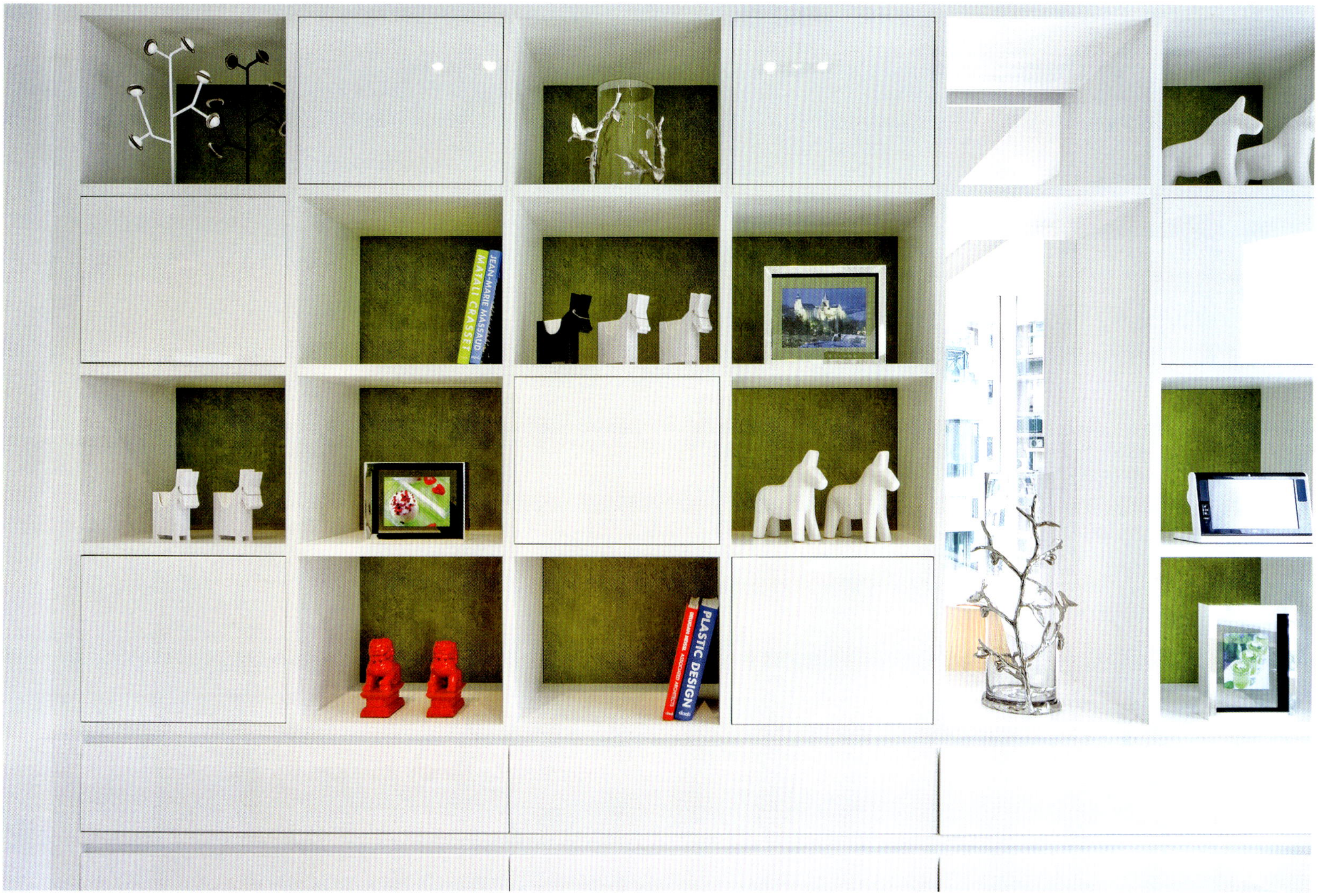
JEAN-MARIE MASSAUD
MATALI CRASSET
PLASTIC DESIGN

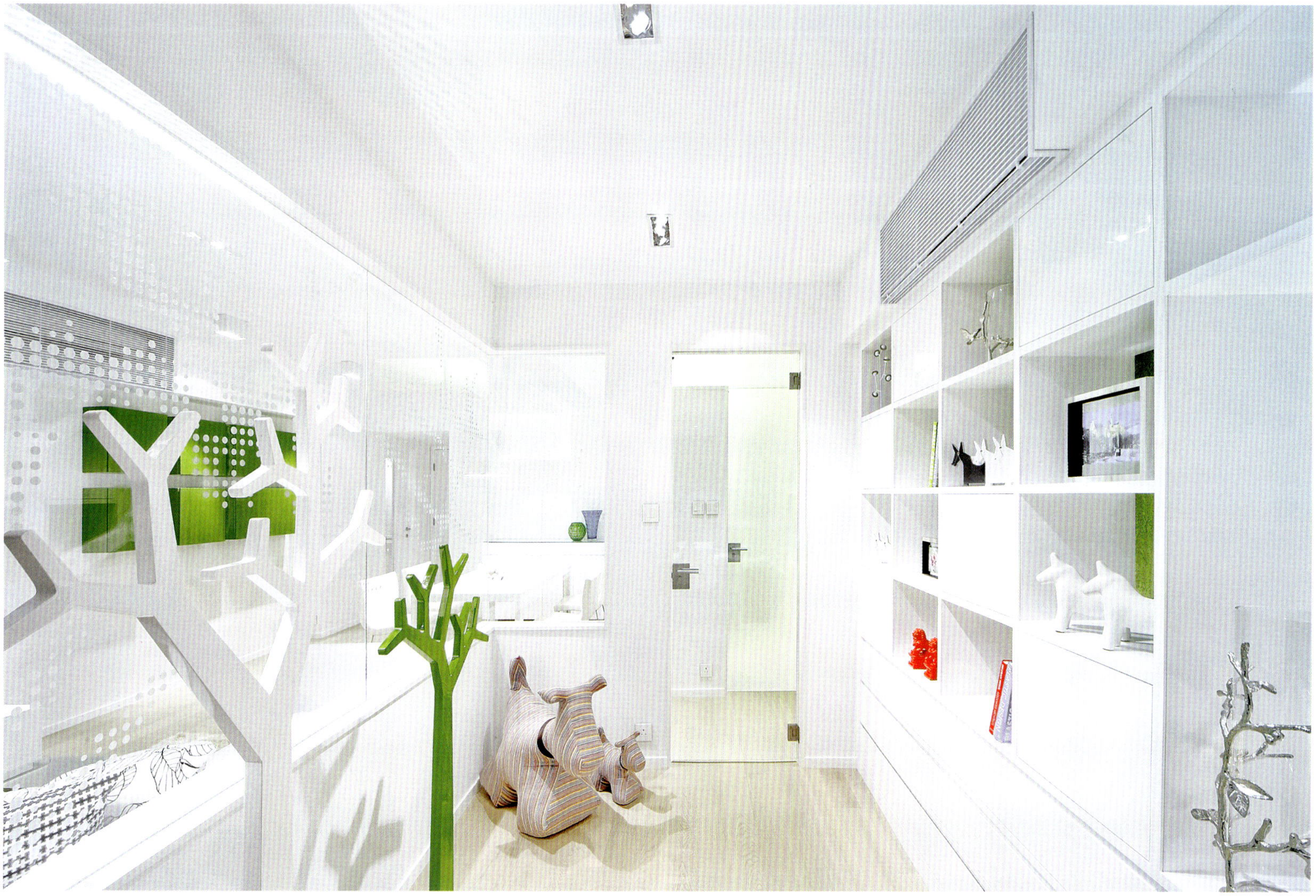

Majestic Height

君临天华

项目地点：福州
项目面积：120 平方米
设计师：陈立风
设计公司：简风设计工作室
主要材料：玻化砖、马赛克、水曲柳面板、烤漆玻璃

The project is Mediterranean style, where the interior door frames, windows, and seat covers are decorated with Mediterranean collocation of blue and white, coupled with shells, wall covered with silver sand, ground with small pebbles, collage mosaics, metal containers and other elements, showing the different-level contrast and combination of blue and white to the limit.

The designer makes some adjustments on the unite, where the garden at the entrance is changed into a dining room, and the arc-curved wall is changed into an arc French window; the wall facing the entrance is opened, so the line of sight from the entrance is transparent and bright and the space is more spacious; the end of corridor is planned to be arc, echoing with the dining room, to resolve the feeling of depression and narrowness from the corridor. The wash station of the guest washroom is placed outside to form wet and dry areas, which is more convenient.

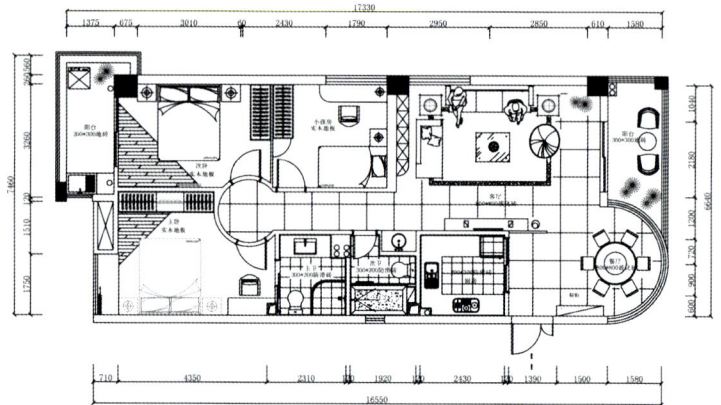

本案采用地中海风格，室内的门框、窗户、椅面都是地中海式的蓝白配色，同时搭配了贝壳、细沙装饰的墙面、小鹅卵石地、拼贴马赛克、金属器皿等元素，将蓝与白不同程度的对比与组合发挥到极致。

设计师对户型做了一些调整：将入门的入户花园改为餐厅，弧形外墙被改为弧形落地窗；正对入口的墙面被打通，使入口的视线通透明亮，空间更为宽敞；将走廊尽头规划为弧形，与餐厅相呼应，化解了直线走廊带来的狭窄压抑的感觉；客卫的洗脸台设置于外面，形成干湿分离，使用上更为方便。

Huang's Residence on Wooden Grid

木栅黄宅

项目地点：台北
项目面积：115.5平方米
设计师：杨焕生
参与设计师：郭士豪、王慧静、王莉莉
设计公司：杨焕生建筑室内设计事务所
主要材料：喷漆白、明镜、大理石
摄影师：刘俊杰

In the project, the designer chooses hidden sliding doors to integrate two originally narrow spaces into a spacious and interflowing space. The design is based on white, and the crystal droplight leaping to the eyes shows out a romantic and elegant atmosphere. In the living room, the sofa background wall is extended and turned over to form the extension of the wall with the French window, so the vision extends to the outdoors. The TV wall is decorated with silver fox marble to show the introverted character, and the ceiling is embellished by multi-level way of delicate carving to add elegant atmosphere into the space. The guest wash room and the dining room are designed with hidden sliding doors, and the reflection of mirror glass is used to make the rooms more spacious. The master bedroom continues the white atmosphere, added with delicate carvings, and the hidden indirect lighting integrates the modern and romantic elegance skillfully.

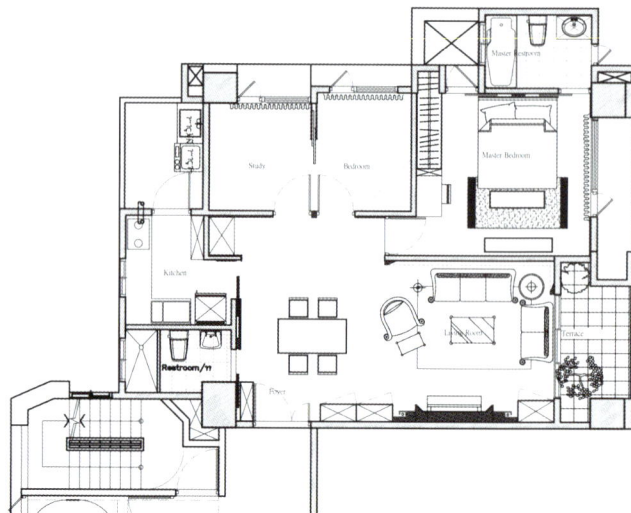

本案的设计师利用隐藏式拉门，将原本狭小的两个空间融合为一个互通、宽敞的空间。设计以白色为基色，映入眼帘的水晶吊灯显现出浪漫与典雅的气息。

客厅沙发背景墙的延伸和转折处理，形成了落地窗墙面的延续，使视野延展到户外。电视墙面运用银狐大理石彰显内敛性格，天花板以经过细腻雕刻的多层次面板装饰，让空间更添优雅的气息。客浴与餐厅运用隐藏式拉门，并利用镜面玻璃反射使空间更显宽阔。主卧室延续白色氛围，更添加细腻的雕刻品，隐藏的间接光源将现代与浪漫优雅巧妙结合。

Banqiao Culture City

板桥文邑

项目地点：台北
项目面积：264平方米
设计师：马健凯
设计公司：界阳&大司室内设计
主要材料：烤漆玻璃、钢琴烤漆、不锈钢、舒然板、发光二极管制成的灯、卡拉拉白大
理石、日本进口壁纸

The designer chooses ingenious collocation to integrate the materials full of modern sense and ratio of lines. The entrance is connected with the interior floor with metal tiles, and the shoe cabinet is coupled with light box to create a unique atmosphere, where the clear visual hierarchy features places with different functions. The sense of pressure at the entrance is feathered through the introduction of mirror. The transparent curved glass is used as the partition between the study and the living room, extending the view of the living room without obstruction and creating a unique effect of deep scene, and the extending indirect lighting enlarge the visual width of the living space. The master bedroom is designed with the collocating way of classic black and white, with exquisite crystal droplight hanging in the corner, interpreting a fashionable and luxurious atmosphere. The changing room is designed with mirror as the walls, with functions of enlarging the space and make-up mirror.

设计师通过巧妙搭配，整合现代感十足的素材与线条比例。由玄关处金属砖衔接室内地面，鞋柜下方搭配灯箱，营造专属氛围，明确的视觉层次规划出不同的功能场所。通过镜面的设置，虚化了入口空间的压迫感。设计以通透的弧形玻璃作为书房分隔墙，使客厅视角延伸无阻，创造出空间独特的景深效果，延伸的间接光源同时放大了客厅的视觉面宽。经典的黑白两色搭配手法、角落吊挂的精致水晶灯诠释出主卧的时尚与奢华氛围。更衣室以明镜为墙，兼具放大空间与整装镜的功能。

Cai's Residence of Beautiful Scene

景美蔡公馆

项目地点：台北
项目面积：110平方米
设计师：虞国纶
设计公司：格纶设计
主要材料：明镜、清玻璃、浅金锋大理石材、白色烤漆、榉木原色地板、得利乳胶漆

The American rural and neo-classical styles are integrated in the project, showing an extraordinary momentum. The living room breaks the traditional thinking with saturated and bright colors to show the warm temperament of the space, and the dining room is designed by concentrating way with eye-catching blue dining chairs and white background tone to produce sharp contrast, strengthening the delicate impression of neo-classical style. After the process of dyed white teak, the old dining table and chairs regain new life, cleverly integrated into the new space. In order to make unified style, the Japanese-style space is designed with large areas of floor glass, where the door is decorated with white glass, and the windows are designed with neat horizontal-stripe patterns, extricating the oriental impression of Japanese style and inheriting the American classical and rural style. In the master bedroom, the red headboard, carved decorative mirror, and sheep-horn-shaped red crystal wall lamp show the vitality of the space.

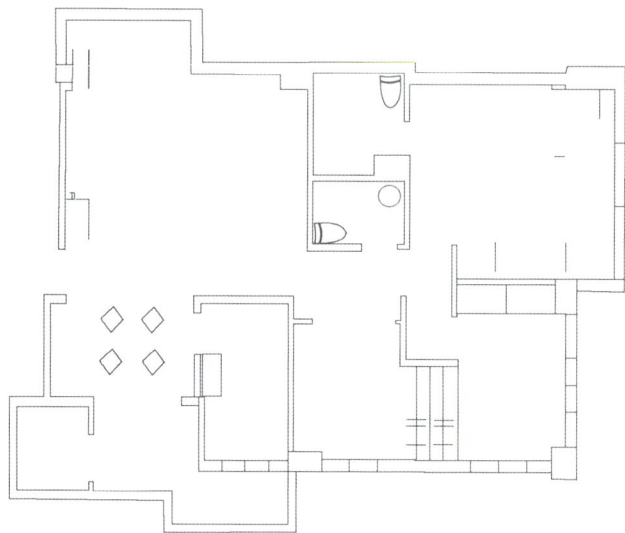

本案融合了美式乡村和新古典风格的精神，展现不凡气势。客厅空间突破传统思维，以饱和明亮的色彩展现空间温暖的气质。餐厅空间以浓缩的方式呈现，以抢眼的蓝色餐椅与白色背景基调形成强烈对比，强化新古典的精致印象。经过柚木染白的过程，旧的餐桌椅重现新生命，并巧妙地融入新的空间中。为了使风格统一，和室采用大面落地玻璃，门片以白色玻璃装饰，落地窗则设计为利落的线板横条纹样式，摆脱和室的东方印象，承袭美式古典及乡村风格。主卧室红色床头板、雕花装饰镜、红色羊角状水晶壁灯，绽放着空间的生命力。

Passion of Off-road Sport

越野豪情

设计师：黄志达
设计公司：黄志达设计师有限公司

Off-road sport is a kind of sport that integrates competitiveness, enjoyment and excitement into one, and further is an attitude that experiences life and outbreaks one's own life. The designer takes the off-road spirit as the dominate spirit, and creates a natural miracle and charming experience everywhere inside the project. No matter the decoration of flowers and grass patterns on the walls, the three-dimensional hanging furnishings and paintings framed by metal, or the cattle hide carpet with black and white mosaic lattices on the floor, all make one feel the harmonious relationship with nature from detailed life.

The designer, in a bold way, uses rich spatial colors to reveal the passionate off-road feelings. Such feelings have already penetrated in every detail of the apartment, and make the residence become a harbor that is more refined, tolerant and possess more love for humanity.

越野是集竞技性、观赏性、刺激性于一体的运动，更是一种体验自然、突破自我的生活态度。设计师以越野精神为主导，在室内处处营造出自然的神奇与迷人之体验。无论是墙面花草纹样的装饰、金属相框的立体挂饰和画作，还是地面黑白格子拼花的牛皮地毯，都使人于点滴生活中感受与自然的和谐相处。

设计师大胆地使用了浓重的空间色彩，体现了炙热的越野情怀。这种精神已经渗透在居室的每一个细节中，使居室成为更儒雅、更宽容、更具有博爱精神的港湾。

Milan Impression, 3#C-2 Baoli Garden

保利花园3#C-2米兰印象

项目地点：长沙
设计师：王奰、王小锋
设计单位：广州尚逸装饰设计有限公司

This project places much emphasis on the employment of classic furniture and accessories, along with the simplified and bright decoration style, the characteristic of Milan, the Fashion city, is thus brought into the interior design. Moreover, treated with the modern technique, the designers create such a living space that is practical and fashionable as well as elegant.

In terms of design, the designers simplify the original complex details of the European style, and combined with the contracted modern design technique. They use elements, which have been simplified but still carry European symbols, to create such a space that possesses not only succinctness but also traditional European class.

In terms of the choice of materials, the floor of the public space chooses smooth-surfaced, light colored stone bricks as the principle covering, and the elevation and ceiling apply the combined use of white painting, solid wood lines and wire frames. Apart from that, the master bedroom and secondary bedroom employ more European-styled classic wall paper with floral patterns, gentle and soft fabric and carpets. As far as the choice of the colors of materials is concerned, it takes light color systems, such as white and cream-color, as the theme; therefore, the succinct and clean color match can naturally set off the exquisiteness of the European-styled furniture and accessories.

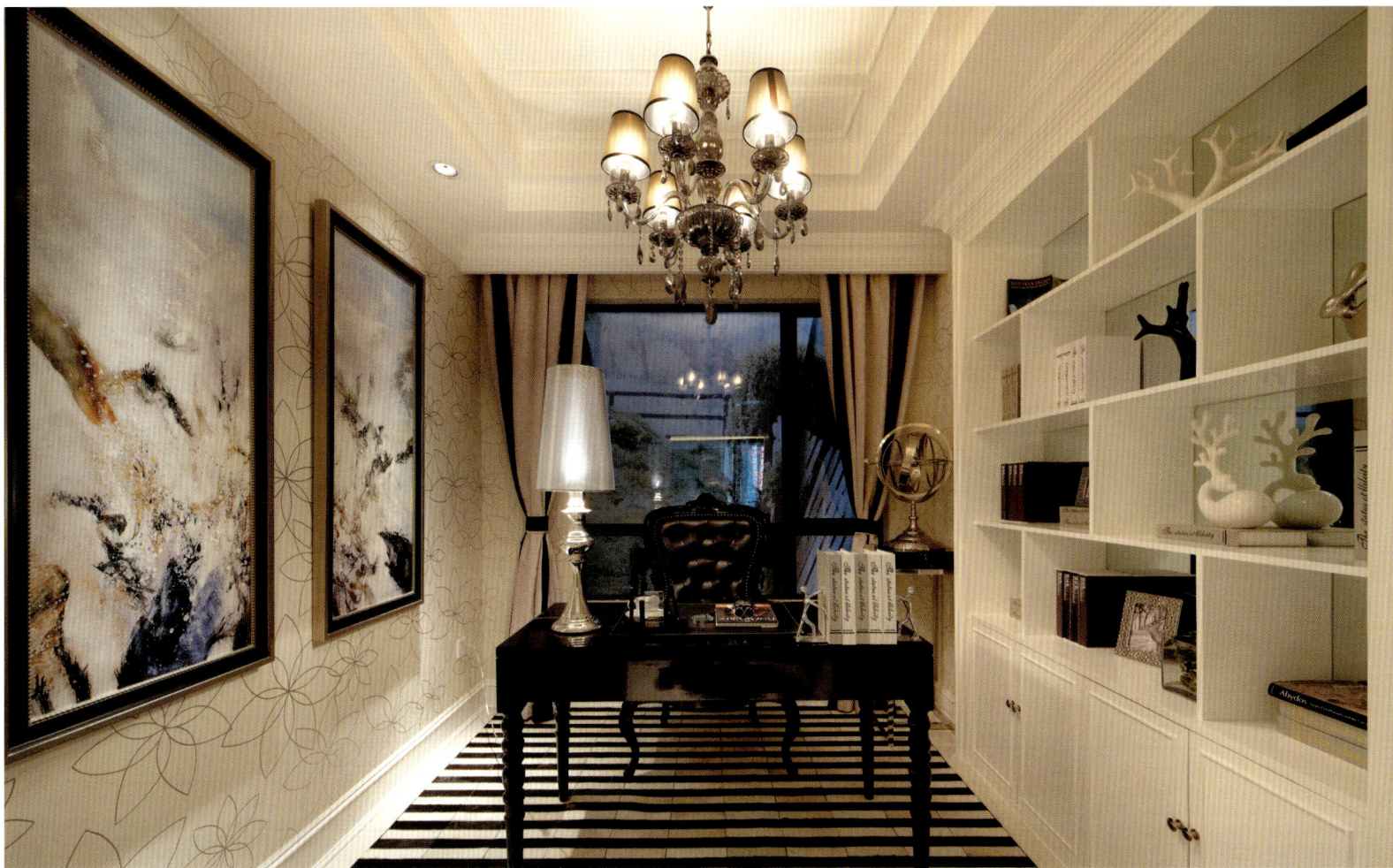

本案着重引用经典的家具、配饰品，配合简洁明亮的装修风格，把米兰这种时尚之都的特点引入室内设计中，用现代的处理手法打造既实用又不失时尚典雅的生活空间。

在设计上，把欧式风格原本繁琐复杂的细节简化，结合简约的现代设计手法，通过简化了却依旧带有欧式象征性的元素来塑造空间，既具有现代简洁的特点，又不失传统欧式风情。

在材质的运用上，公共空间地面以光面的浅色系石砖为主，立面与天花上是白色涂料及实木线、线框的结合运用，而在主卧、次卧等空间中则多采用欧式经典的饰花纹样墙纸、温和柔软的布料、地毯等。在材质色彩的选择上，以白色、米黄等浅色系为主，简洁、干净的色彩搭配很自然地衬托出欧式家具与配饰品的精致。

Hangzhou Zhongbei Garden, Phase Two, Show Flat Type B

杭州中北花园二期B户型样板间

项目地点：杭州
项目面积：167平方米
设计师：徐少娴
设计公司：Gotomaikan International Limited

The project is neo-classical style, and it shows out the designer's exquisite design concept from the whole to the details.

In the living room, the wall is decorated with wood, and moldings are simple and full of details. The bright green sofa is the bright spot in the space, particularly eye-catching among the wood color, and the luxurious but not complicated crystal droplight perfects the noble atmosphere of the living space. The brown TV background wall with twill lines is coupled with the painted beige walls on both sides, which is harmonious. In the selection of furniture, the designer pays attention to keep along with the theme of the design on color and form, and introduces a large number of symmetric methods to make the interior atmosphere more dignified and elegant.

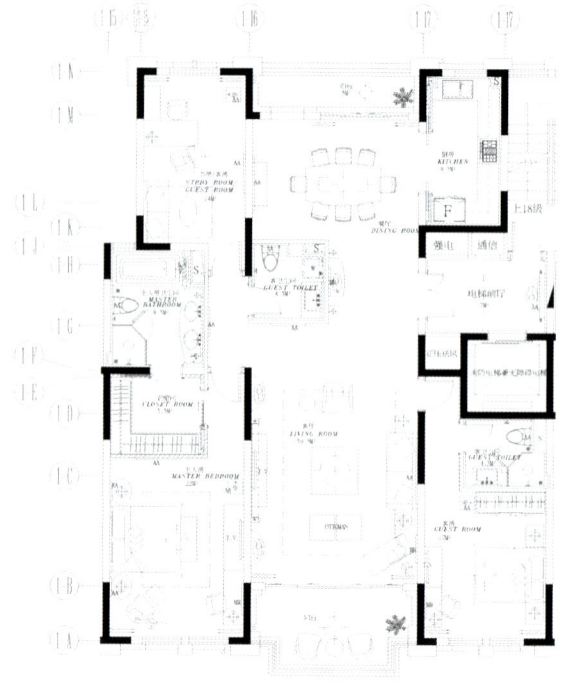

本案采用新古典的风格，从整体到细节均体现出了设计师精致到位的设计理念。

客厅墙面采用木质处理，线脚简单却充满细节感。鲜明的绿色沙发是空间中的亮点，在空间中的木质色彩中格外抢眼。奢华但并不繁复的水晶吊灯完善了客厅空间的华贵气氛。棕色斜纹电视背景墙与两侧米黄色的漆面墙搭配，色彩和谐。在家具的选择方面，色彩和形式上都注重跟设计的主题风格相统一。大量对称手法的运用使室内气氛更为端庄和高雅。

Bright Leafage

发光的叶子

项目地点：台北
项目面积：22平方米
设计师：张嘉芳
设计公司：真观空间设计有限公司
主要材料：白橡木染黑、天然玉石、黑镜、雾面石英砖、染色木皮、灰镜、超耐磨地板

The project is a renovation of an old house, and the designer changes the original narrow three-room house into a resilient "1+1" house, in addition to the bright and spacious living room and open-style dining and cooking room, a master bedroom with complete functions and exclusive bathroom and a multi-function study room with beauty and flexibility are added into the space. The designer plans out a hidden storage intimately, increasing the interior storing space.

The whole style highlights a moderate blank, avoiding excessive decorating elements. The original balcony area is designed to be the entrance with bright French window instead of parapet, and a light white shoe cabinet is customized in the face of the door. The bright leafage on the top of the corridor meets the function of lighting, modifying the beams.

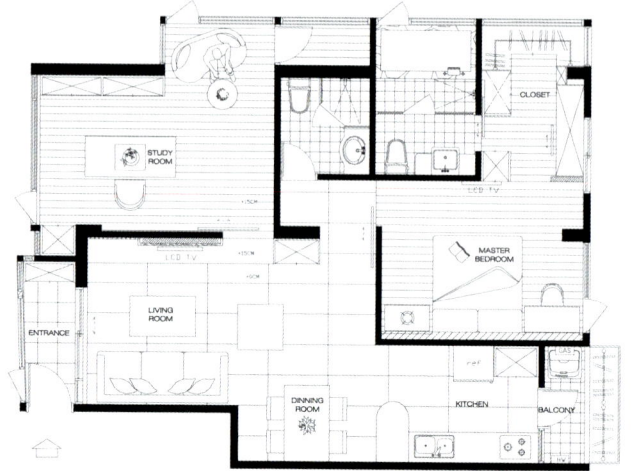

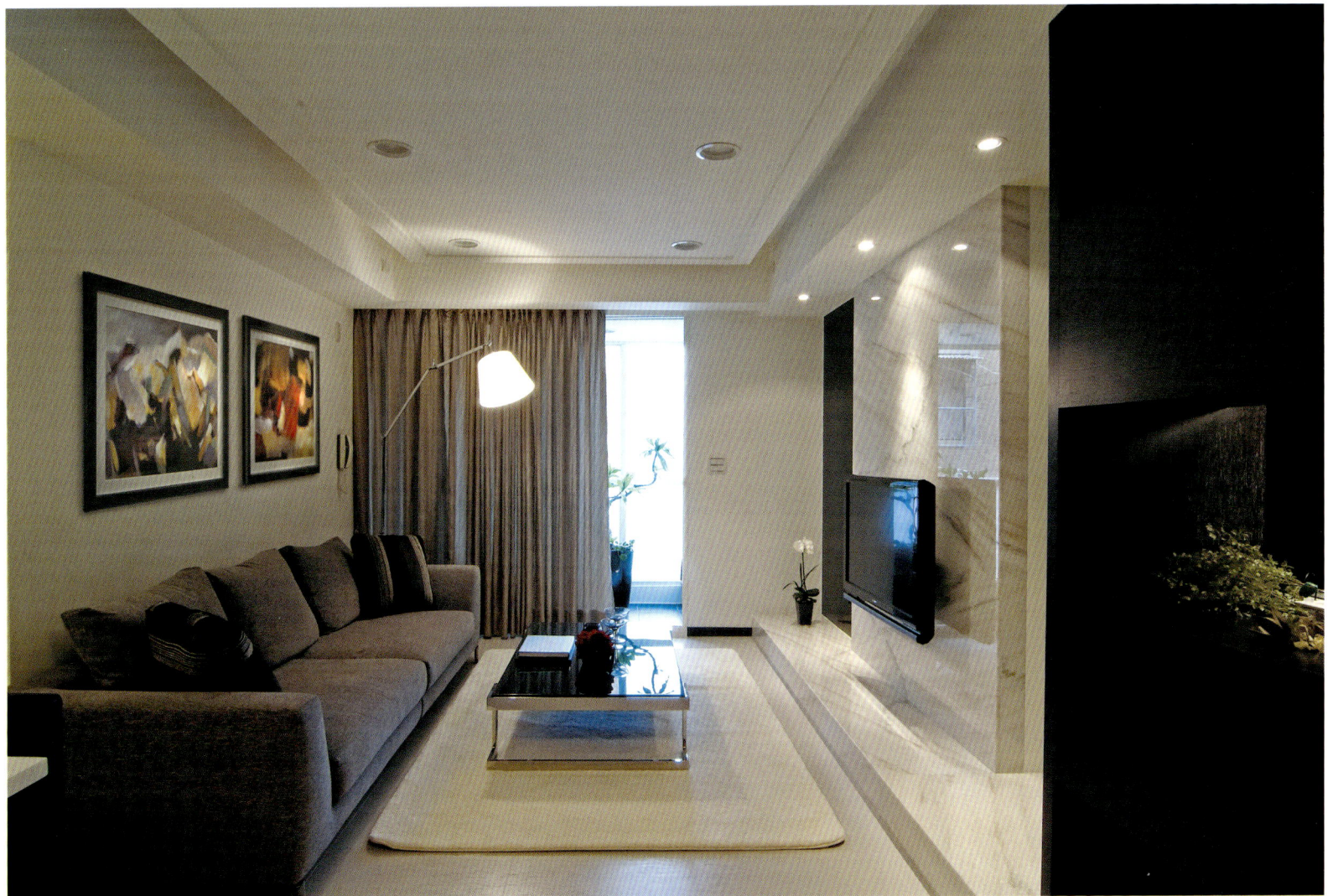

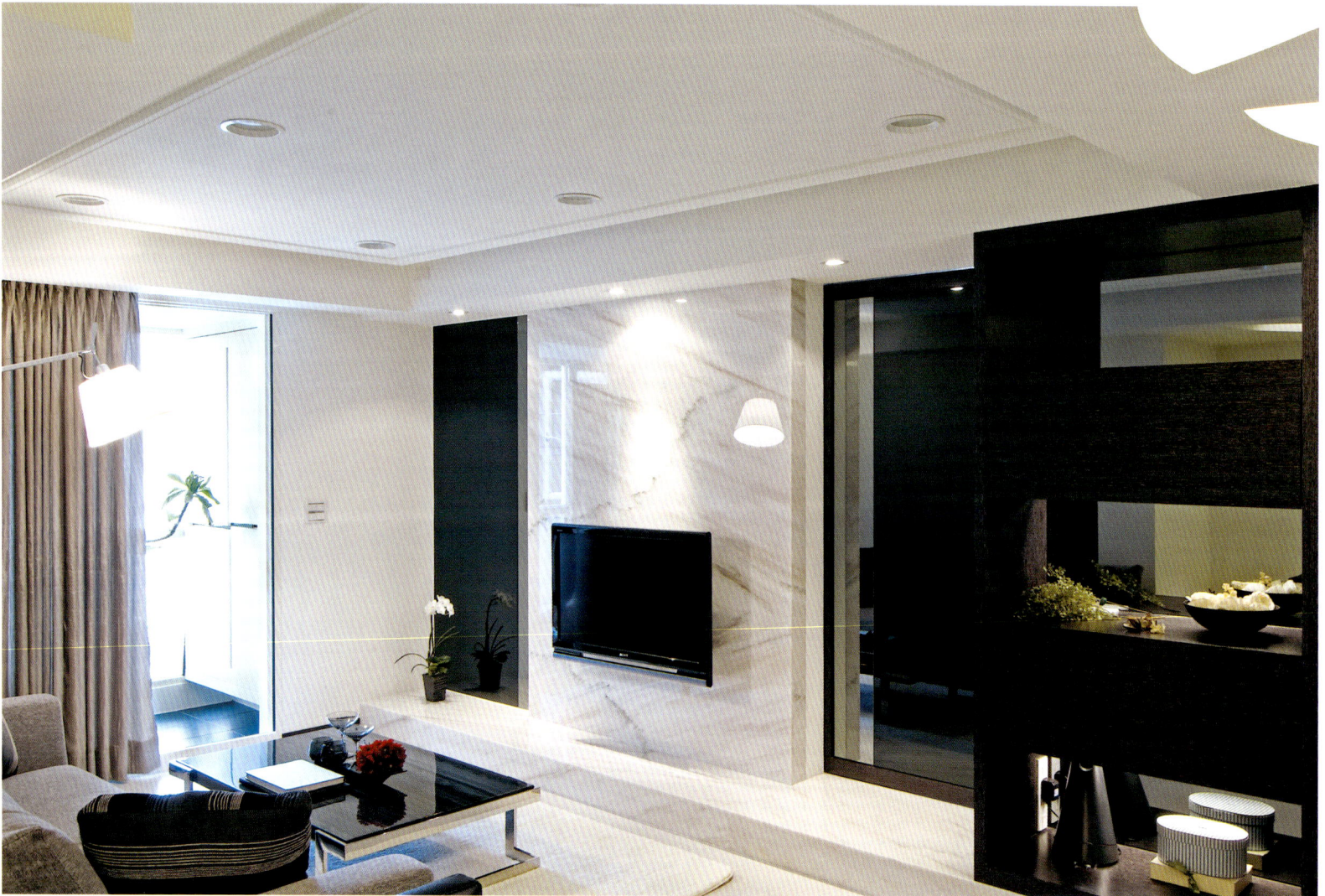

此方案为旧宅改造设计，设计师将原有窄小的三房户型，改成实用又富于弹性的"1+1"户型，除了明亮、宽敞的客厅与开放餐厨区外，还增设了功能完备的主卧室、专属浴室以及一间兼具美观与灵活性的多功能书房。设计师更贴心地规划出了一处隐密的储藏间，增加了室内收纳的空间。

整体风格强调适度的留白，避免过多的装饰元素。设计师将原来的阳台区域规划成玄关，用明亮的落地窗取代旧的女儿墙，并在大门正面特别制作轻盈的白色鞋柜。走廊上方的发光叶子不仅能满足照明的功能，更兼具修饰横梁的意义。

Delicate Space, Revel in Fragrance

细腻空间　暗香盈袖

项目地点：福州
项目面积：187平方米
设计师：吴毅鹏、周颖
设计公司：福州臻美空间设计事务所
摄影师：施凯

In the project, the designers choose bright Mediterranean style and take use of white walls, blue sofa, golden yellow tea table, arched doors and windows and other elements to make the interior atmosphere clean and comfortable. The watercolor on the wall depicts golden sunflowers under the blue sky, setting off the happy and warm atmosphere of the family. Looking from the living room, you can see arches everywhere, so each part of the space is connected with each other and the interior space is brighter and more spacious. Through this kind of treatment, you can see the wall painting of sunflowers from the arch of the dining room. The clean and bright desk and hand-made white lime plaster are original and true, and through the deal of details, the deep style of Mediterranean is introduced into every corner of the space.

　　本案设计师采用了明媚的地中海风格，运用白色的墙面、蓝色的沙发、金黄色的茶几、拱形的门窗等元素，使室内氛围清新而舒适。墙面上的水彩画描绘着蓝天白云下金灿灿的向日葵，烘托了家庭幸福、温馨的氛围。从客厅看去，随处可见的拱门使各个部分的空间相互流通，使室内更明亮宽敞。使人能够透过餐厅的拱门看到客厅的向日葵壁画。简洁明亮的书桌、手做的白色灰质抹面，原始而真实。通过设计师的细节处理，地中海的浓浓风情被引入到室内的各个角落。

Zhan's House, Taipei

台北詹宅

项目地点：台北
项目面积：198平方米
设计师：孙铭逸
设计公司：大宣设计工程公司
主要材料：银狐大理石、灰姑娘大理石、紫檀实木地板、烤漆玻璃、施华
洛世奇水晶灯、皮革、铁件、不锈钢、黑镜
摄影师：小雄梁彦

The owner of the project is a young couple, and they expect to have one child in the future, so the original 198m² space with four rooms is re-planned into a simple space with a master bedroom and a child room, opening the width of the space and creating more comfortable and free atmosphere without pressure. From the entrance, along the living room and dining room to the study, the open design without separation forms the interconnected space vocabulary. In the space planning, addition to dividing the private bedroom into two rooms, the public space is mainly open and flowing design without boundaries, and the elements of blocks and lines, and changes of materials are used to define the areas in the space, so through the layout and disposition of the design, the couple can enjoy close contact and interaction without pressure.

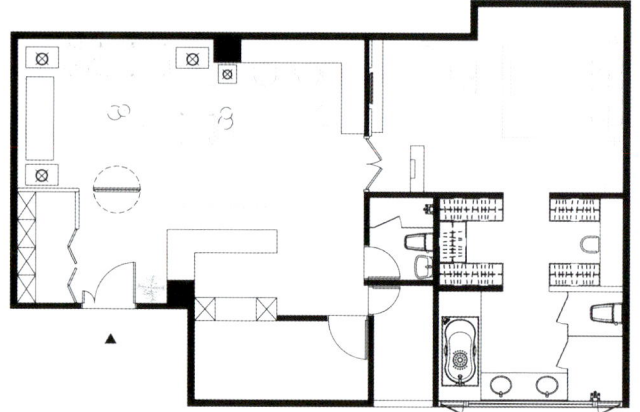

本案屋主是一对年轻夫妇,预计在将来生一个孩子,因此将原本四室的198平方米的空间重新规划为单纯的主卧室与儿童房,拉开空间广度,营造更舒适无压的自在气息。从玄关进入,沿着客厅和餐厅来到书房,开阔无隔间的设计构成了相互连接的空间语汇。在空间的规划上,设计师将私密的卧室规划为两室,公共空间尽量以开阔流动的无界线设计为主,运用块体、线条元素以及材质的变化作为空间中的区域界定,通过设计的铺陈与配置,让屋主夫妇能获得紧密却无压的互动联系。

Spring of Wan Tai

万泰春天

项目面积：340平方米
设计师：彭东生
设计公司：汕头市天顺祥设计有限公司

The exquisite workmanship is an important feature of American style. In this project, it can be considered as the resonance of thoughts of both the middle-aged designer and the owner. The experience of life and the deposit of thoughts have found their expression in the excellent making-old technique on the panel of the same materials: the frames, doors and grills, from outside to inside. From the living room to the dining room, and from the corridor to the rooms, not a single design makes a display of showing off; instead, they take on a built-in nobleness. The designer helps people feel that what needs to be duplicated is not the pure American style, but the life attitude that focuses on the essence of life.

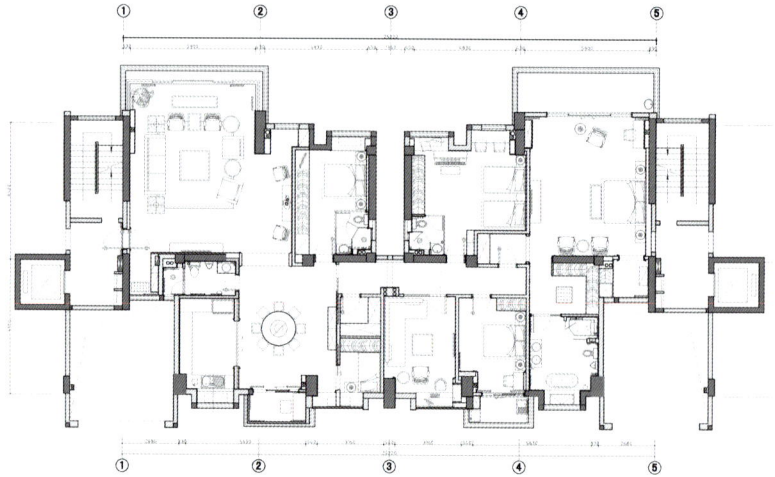

精湛的做工，是美式风格的重要特点，这套作品可以看做是中年设计师和中年业主的思想共鸣。生活的历练和思想的沉淀，化成了这套作品从外到内的框架、房门、格栅等所有统一材质的面板上的精良的做旧工艺。从客厅到餐厅，从走廊到房间，所有的设计毫无炫耀，自有一种骨子里的尊贵，设计师让我们感悟到，需要复刻的，不是单纯的美式家居风格，而是注重内涵的生活态度。

Green Flash, Residence L

绿光L宅

项目地点：台北
项目面积：148.5平方米
设计师：唐忠汉
设计公司：近境制作设计有限公司
主要材料：石皮、柚木实木、杉木实木、黑铁

A main stone material wall as the distinction between inside and outside spaces starts the planning of the space, showing the open characteristic of the interior space. The designer uses the moving sliding doors in the walls to connect or divide each space, blurring the definition of spaces and redefining the attribute of spaces. The couch platform in the space is another important element, connecting the two places with similar shape but different functions. Combined with the planting near the window, the platform features another function of connecting the spaces.

Through the use of nature, light, and materials, the designer starts from the most primitive characters of the space and hopes to refine the most concise essence of space through this way.

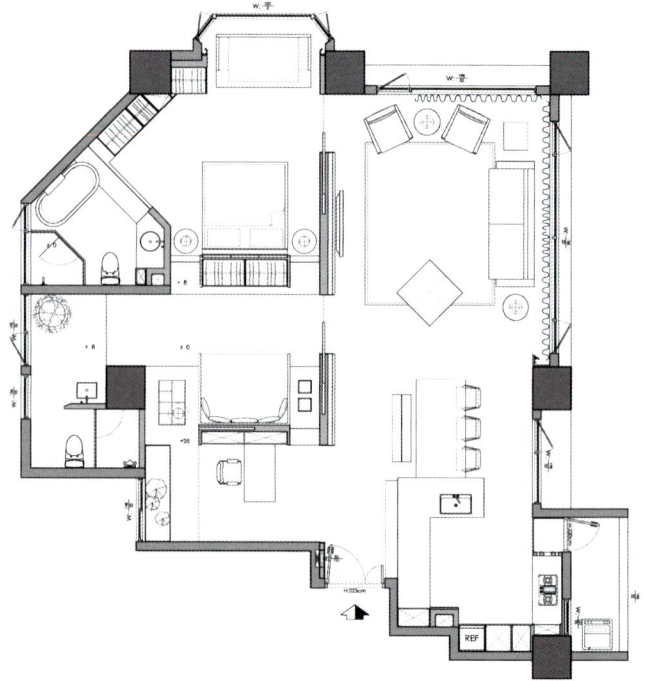

　　对空间的规划由一道主要的石材墙面展开，该墙面区分了内外空间，也体现了室内空间开放的特质。设计师利用墙面中移动的拉门将各空间连结或区分，模糊了空间的界线，重新定义了空间的属性。空间中的卧榻平台作为另一个重要元素，连接着形态相似、功能不同的两个场所，结合窗边的植栽，平台的形态具备了连接空间的另一层意义。

　　设计师通过对自然、光线与材质的运用，从空间中最原始的特征着手，希望通过此种手法淬炼出最精简的空间本质。

Venture on the Coral Island

珊瑚岛奇遇记

项目地点：深圳
项目面积：88平方米
设计师：导火牛
主要材料：通体砖、外墙防水漆、马赛克、原木、东北红杉
摄影师：导火牛

The designer has captured different "inspirational segments" in the process of his design, including magic, surrealistic design style, the comic strips of Mystery Island, works of Henri Matisse who is the special genius artist, the candle drop lights in Grimm's Fairy Tales, and the tawdry picture color matching of Zuma game. The designer has turned the design process into a fun treasure hunt. He makes the sofa area a coral island with blossoming flowers, transforms the television wall into a sailing boat floating in the living room, with one sector of the hanging cabinet the form of rolling waves. Through his design, the designer has put different "inspirational segments" into an enchanting treasure map, which sails toward the coral island at heart.

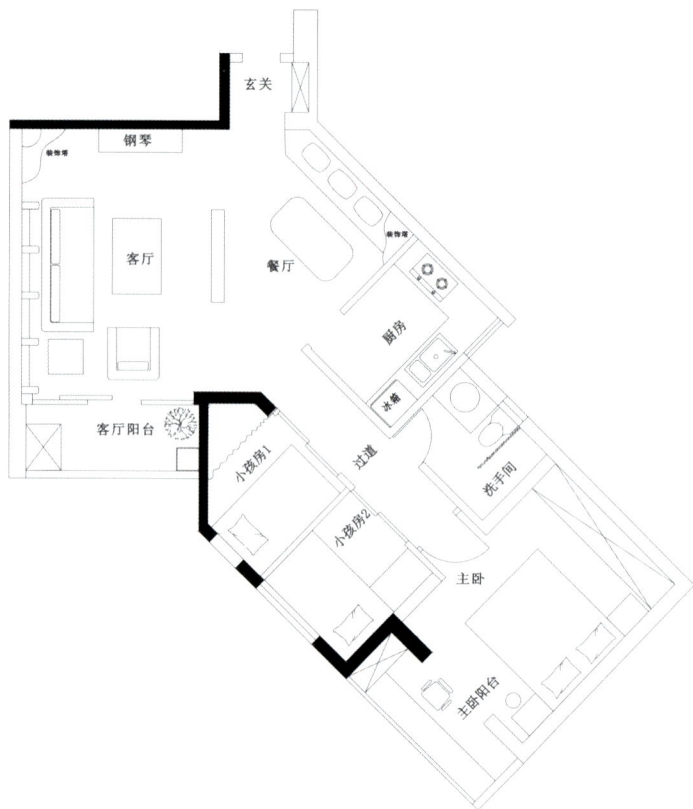

設計師在設計過程中捕捉到了不同的"灵感碎片"，其中包括魔术、超现实主义的设计风格、连环画《神秘岛》、鬼才画家亨利·马蒂斯的作品、格林童话中的蜡烛吊灯、祖玛游戏非常华丽的画面颜色搭配等。设计师把设计过程变成有趣的寻宝游戏，将沙发区域营造成春暖花开的珊瑚岛，将电视墙变成一艘漂浮在客厅中的帆船，将厨柜的一节吊柜营造成起伏的海浪。设计师通过设计将不同的"灵感碎片"拼成了一幅美丽的藏宝图，驶向了心中的珊瑚岛。

Still Time

静止的时光

项目地点：台北
项目面积：120平方米
设计师：俞佳宏
设计公司：尚艺室内设计有限公司
摄影师：岑修贤

Still time means letting the sunlight stop and making the time congeal. By giving this concept, the designer aims at, with his design, creating a space that blurs the concept of time and can only let one feel the presence of human beings.

The succinct screen chooses a dark colored and neat style, the iron window grilles are modest yet complicated, which enhance the perception of depth of the foyer and dining room. The open space integrates the household elements of the living room, study room and dining room in an unreasonable way. The designer uses a low and short television wall to separate the living room, and creates a tranquil office corner. Supplemented with the sophisticated and sober materials, dark colored tape strys full of Zan feelings, it creates an artistic conception where time stays still.

　　静止的时光意味着让阳光不变、令时间凝结。设计师提出这一概念旨在通过设计创造一种模糊了时间概念的、只能感受到人主体存在的空间。

　　简洁的屏风采用深色利落的形式，铁件窗花低调而繁复，在视觉上增加了玄关及餐厅的景深感。开放式的空间将客厅、书房、餐厅等生活元素合理整合。一道低矮的电视墙将客厅的部分空间隔开，形成了一个宁谧的办公角落，配合以内敛沉稳的材质、禅味十足的深色挂毯，营造出时间在此停留的意境。

Wenzhi Metropolitan

文质大都会

项目地点：台北
项目面积：115.5平方米
设计师：张嘉芳
设计公司：真观空间设计有限公司
主要材料：进口大理石、黑镜玻璃、仿古刷纹木地板、进口壁纸、天然木皮

Through the designer's partial adjustment, the original four-room pattern is changed into a three-room pattern with a study room of aesthetic function and mobility. The interior space avoids unnecessary decorative elements, and the overall style is open and free. The facade of the entrance is designed with a line of tall grille screens to protect the privacy of the living room, and the open living room of momentum is filled with rich changes of lights and shadows. The French window leads the outdoor beauty into the space, setting off the neat TV wall. The lines with delicate texture of selected natural stone create the central wall and the floor table at the bottom of the wall. The CD cabinet corresponding to both sides is decorated with wood-woven appearance to show the perfect balance on vision and neutralize moderately the cold feeling of stone.

原有的四室经过设计师的局部流线调整，变成了三室及一间兼具美观与机动性的书房。室内空间避免多余的装饰元素，整体风格开阔而自由。

玄关正面安排一列高挑的格栅屏风，保护客厅对外应有的隐私性。开阔而有气势的客厅充满了丰富的光影变化。落地窗将窗外的美景引入室内，衬托利落的电视主墙。精选天然石材细致的线条纹理，打造中央墙体与下端连结的落地式台面。两侧对应的光碟柜以实木编织的外观展现完美的视觉平衡，同时在视觉上适度中和了石材的冰冷感。

Casa Mataró

马他亚住宅

项目地点：巴塞罗那
项目面积：190平方米
设计师：Elia Felices interiorismo
摄影师：Rafael Vargas

Behind an old monumental facade creates a cosy home, and at the same time an art gallery, located on Mataró (Barcelona).

On the first floor there is the receiver. There is a couple of restored rocker arm, a Baroque indigo blue upholstered sofa and a table center with mirror polished finish. Sober walls in grey, "le nouveau noir", give the space a cool feeling. The dining table center with glass envelope slightly oval and central foot steel stainless mirror polished finish creates a contrast with the chairs of Baroque mouldings, satin and floral motifs. Black and white curtains give a scenic nature for the room. For the kitchen a worktop of white stone, Central Island with Bell suspended in steel and black lacquered furniture.

In the upper Hall two cubes mirrored as nightstands, two restored Baroque moldings and a sofa with organic mouldings with black velvet, time, gramophones, cushions chairs, lamp on a console-style Louis XIV lacquered in black and a monumental portrait reflected before the great mirror of golden Baroque molding. The mixture of periods and styles give personality to the house.

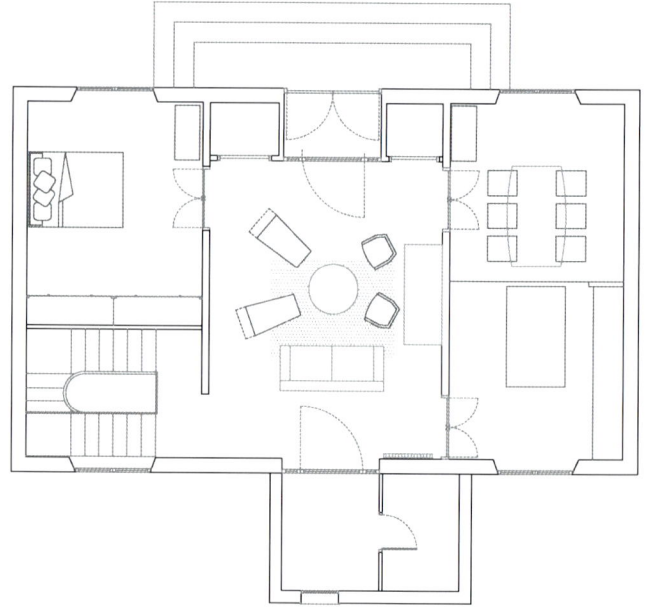

本案位于巴塞罗那老纪念碑后面，这里不仅是一个舒适的家，同时也是一个艺术画廊。

一楼是接待室，包括一对修复的摇臂、巴洛克靛蓝软垫沙发以及抛光镜面台中心。"新黑色风格"灰色墙面使空间显得冷静。餐桌中心饰有玻璃罩面，配上环绕中心的椭圆形不锈钢抛光镜面，使巴洛克风格的椅子线条和绸缎花图案形成强烈对比。黑白色窗帘赋予空间优美的自然风景。厨房设有白石台面，贝尔中心台悬浮于钢质黑漆家具中。

在二楼大厅中，两个相互映照的立方体成为床头小桌，这里还包括两个修复的巴洛克造型、有机线条黑绒沙发、时钟、唱机、带靠垫椅子、控制台式路易十四黑漆灯饰以及金色巴洛克镜面反射出的巨大画像。时代和风格的混搭增添了房屋的个性。

Coral Pavilion

珊瑚阁

项目地点：香港
项目面积：约128平方米
设计师：郑勇威
设计公司：爱家设计有限公司
主要材料：索白色橡木、哑面地砖、木纹地砖、卡撒灰云石、灰镜、不锈钢

The design of this project is led by simple style, with the tone colors of black, white, and gray. In the living room, the TV wall is simply decorated with Casa gray marble, and the horizontal stripes of stone material are steady and elegant. In the dining room, the droplight is specially customized to set off the elegant and gorgeous temperament of the interior space. The wall between the kitchen and the living room is decorated with glass tiles with concave-convex patterns, coupled with double-opened door of glass texture, so the living room and the kitchen are connected with each other on vision. The cupboard is decorated with stainless steel color coating board to highlight the neat character of the space. The partition between the master bedroom and the master wash room is deep gray glass, creating a hazy and intimate sense of space.

本案的室内以简约设计风格为主导，以黑、白、灰为主色调。客厅的电视墙采用卡撒灰云石做简洁装饰，石材的横向条纹稳重而高雅。餐厅的吊灯是特别设计订制的，烘托出室内高雅而华丽的气质。厨房与客厅之间的墙面以凹凸玻璃图案砖装饰，配合玻璃质感的双开门，使客厅与厨房的空间在视觉上形成贯通。厨柜以不锈钢色胶板面装饰，突显空间的利落性格。主卧与主卫之间为深灰色的隔墙，采用了玻璃的材质，营造了朦胧、私密的空间感。

Hangzhou Zhongbei Garden, Phase Two, Show Flat Type D

杭州中北花园二期D户型样板间

项目地点：杭州
项目面积：177平方米
设计师：徐少娴
设计公司：Gotomaikan International Limited

In the design of this project, the designer chooses low-key and gorgeous colors and materials, fully demonstrating the noble temperament through strong expression of metal texture, elegant and imaginative extension space, elegant style, and fine carvings. Metallic colors feature elegance and gorgeousness that can not be replaced by any other colors, and golden sheen is used to decorate the house to render the home flavor greatly. The deal of design details avoids the vulgar feeling, at the same time right to highlight the noble character of the living space. In the space, the designer chooses pure color furniture with delicate details on the design and lines, and the choice of wallpaper and picture frames and other accessories strives to be the same color and similar texture. The design with strong sense of geometry is introduced, and the designer pays attention to the smooth lines of the space.

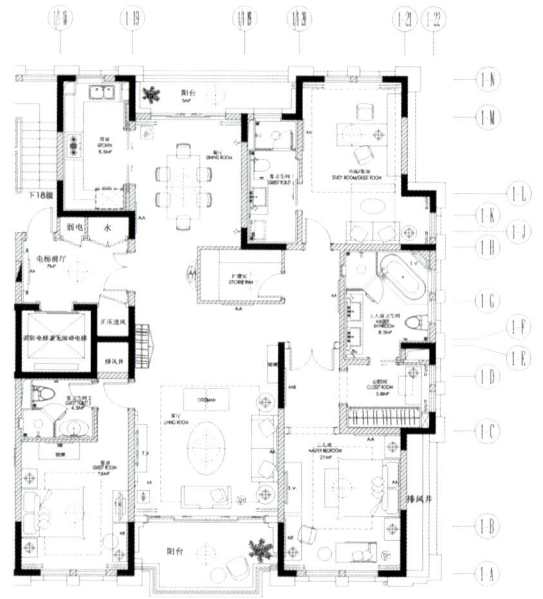

本案在设计中采用低调却华丽的色彩和材质，利用金属质感极强的表现力、高贵与充满想象的外延空间、雅致的造型、细致的雕刻，将高贵气质展露无余。金属色彩，有其他任何色彩都无法替代的高贵与华丽，用金色光泽来装饰住宅，可以极力渲染这种居家风情。通过设计细节的处理，避免了媚俗的感觉，同时恰到好处地彰显出居室的尊贵性格。室内选用了设计与线条都极富细节感的纯色家具，在壁纸、画框及配饰的选择上，尽量以同色系、相似质感为主，而且尽量选择几何感强的设计，注重空间的线条流畅。

Zhubei Zhang's Residence

竹北张公馆

项目地点：中国台湾
项目面积：106平方米
设计师：张巧慧
设计公司：春雨时尚空间设计
主要材料：烤漆玻璃、雾面玻璃、茶镜、烤漆、超耐磨地板、全热交换器

At the entrance, the original solid wall is designed with wooden wall, connecting to the matte glass screen. The lower take-out part of the shoes cabinet not only can be used as dressing chair, but also can be moved into the living room as a tea table. In the living room, the short cabinet is replaced by a curved wall, and the light green TV wall is decorated with wallpaper of lines, simplifying the visual moving lines. On one side, the groove displaying cabinet is coupled with tea-mirror wall, echoing with the mirror above the bookcase in the dining room. Taking the sofa as center, the ceiling is heightened from the center to the living room and dining room, level by level. In order to reduce the sense of pressure, the designer specially chooses the low-back sofa to broaden the vision in the space. In entertaining functions, the design meets the need of children, and the audio-visual equipments used by adults are unified in the half-opening high cabinet.

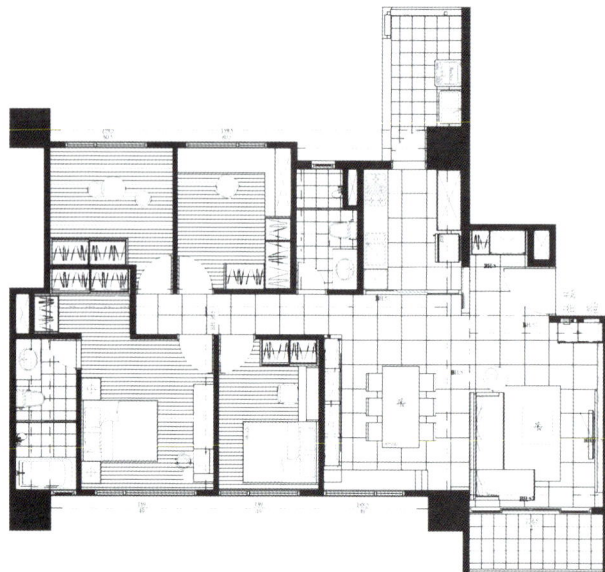

　　设计师在玄关原有实墙的基础上，加做木墙面以衔接雾面玻璃屏风。鞋柜下层抽拉式柜子不仅可作为穿鞋椅使用，还可移至客厅作为茶几。客厅省略了矮柜设计，改以弧形墙面，粉绿的电视墙面搭配线条壁纸，简化了视觉动线。一侧凹槽展示柜搭配茶镜壁面，与餐厅书柜上方的镜面形成呼应。天花板以沙发为中心向客厅、餐厅两侧做出层次性的升高处理。为了减少压迫感，设计师特别选用低背沙发，使空间视野开阔。娱乐功能中除满足小朋友的需求外，成人所使用的视听设备则统一收纳于半开放的高柜中。

Mix and Match with Pure Technology

混血混搭纯科技

项目地点：台北
项目面积：142平方米
设计师：江先立
设计公司：佶舍室内设计
主要材料：大理石、不锈钢砖、特殊砖、烤漆玻璃、
西班牙人造石、木地板砖、灰镜、进口壁纸

The design incorporates diverse international materials, creating a mix and match home with international perspective, and the re-designed and re-planned pattern with two bedrooms and three washrooms endows the old house with luxurious momentum. The designer omits the heavy wood, and the stylish screen at the entrance with penetration of circles and lines creates the light degree of the space. In the living room, the TV wall with rock texture, TV cabinet with black mirror, Mandailing wooden display cabinet, the three functional sections show the continuity of one stroke, echoing with the indirect lighting lines on the sofa background wall with soft lighting. The kitchen is designed with technology to create the taste of home with single product of high texture. The master bedroom is coupled with a washroom pulling the outdoor scene, and FRP thick film is used as water leak proof, which is lasting and durable. The ceiling is decorated mainly with aluminum material, avoiding leaving water stains.

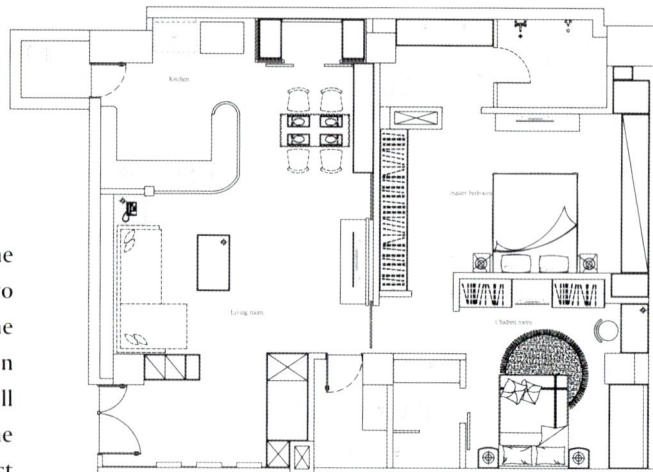

该设计融合了多元国际素材，以国际观打造混搭家居。重新设计规划后的两房三卫格局让旧宅有了豪宅气势。设计师省略掉沉重的木作，玄关处时尚的屏风以圆与线的穿透营造出空间的轻盈度。客厅岩石质感的电视墙、黑镜机柜、曼特宁木作展示柜，三种功能家居展现出一气呵成的连续性，呼应沙发背景墙间接光源的线条，晕化出光源柔美的色调。用科技打造的厨房以高质感单品营造出居家品味。主卧大面引景卫浴采用纤维增强复合塑料厚膜进行防漏水处理，持久而耐用。天花板以铝质材料为主，避免留下水渍。

Blue-white Breath

蓝白色的呼吸

项目地点：成都
项目面积：140平方米
设计师：张静、廖志强
设计公司：之境室内设计事务所

The whole project has taken factors like low-carbon into full consideration, in which it does not adopt too many decorative elements and makes each functional space fully used through design, reaching the integration of energy and efficiency.

The craft combination of architecture and natural environment makes one feel that these apartments seem coming from the underwater world. Opening the small fence of the indoor-garden, what leaps to your eyes is the perfect combination of blue and white, making one feel the infinite imaginary space. The choice of the furniture is stretched along with the theme of ocean, striving for a unified whole. The Mediterranean styled residences have walls painted in colorful paintings and outstanding decorations, which reveal the bright tinge of sky and ocean.

The designer has tried some bold attempts in the treatment of some elements. For example, in terms of the choice of the kitchen tiles, some yellow tones are interspersed.

该项目整体设计中充分考虑了低碳环保等因素，不采用过多的装饰性元素，通过设计使每个功能空间都得到充分的利用，达到能效合一。

建筑与自然环境的巧妙融合，让人觉得这些房屋仿佛来自海底世界。推开入户花园的小栅栏，看到的是蓝色同白色的完美结合，使人感受到无限的想象空间。家具的选择也跟随着海洋的主题展开，力求达到整体统一。地中海风格的住宅通过彩色墙面和醒目的装饰，体现了天空和海洋明快的色调。

设计师在部分元素的处理上进行了一些大胆的尝试。例如，在厨房同卫生间瓷砖的选择上，穿插了一些黄色调。

Golden Glamour Maple Fashion —Kapok Family

金辉枫尚——木棉家

项目地点：福州
项目面积：105平方米
设计师：陈温斌
设计公司：玄风设计工作室
主要材料：杉木、木纹砖、墙纸

The design of this project has employed a modern technique, creating a succinct and comfortable spatial atmosphere. The interior space uses the off-white color as the main tone, and with the matching of off-white and pure white, the whole space is filled with a cozy atmosphere. The wood-grained bricks and the beige cream-colored wall paper bring a sententious visual effect, and the hollowed out partition made of China Fir Wood makes the whole space full of fresh and natural feelings. The open kitchen makes people feel comfortable and naturally while dining here. The pure white embossing suspended ceiling applies natural flowers and grass patterns that are similar to the hollowed partition. As a consequence, the whole project reaches its harmony in its colors and rich in changes in details.

The off-white sofa matches perfectly with the whole space, so that brings out the best in each other, which makes the whole space appeal more harmonious.

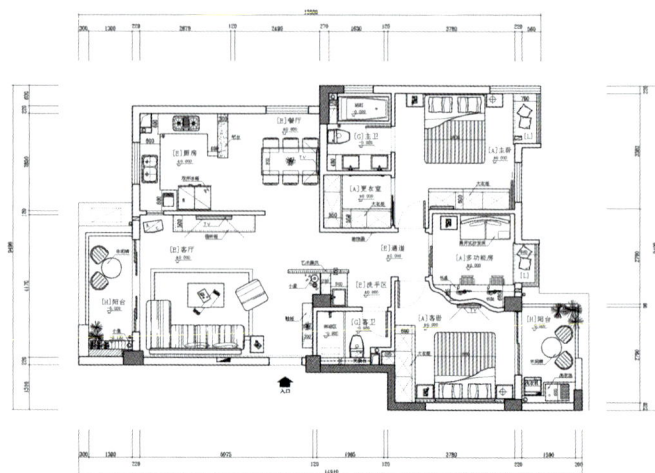

　　本案的设计使用了现代手法，营造了简洁舒适的空间氛围。室内采用了米黄色作为主色调，通过米黄色与纯白色的搭配，使房间充满温馨的气氛。木纹砖、米色墙纸给人简洁的视觉感受，杉木镂空隔断使整个空间清新自然。开放式的餐厅，使人在用餐的时候感觉到舒适、自然。纯白色的浮雕吊顶采用了与镂空隔断相似的自然花草纹路，色彩方面和谐统一，细节方面富于变化。

　　米黄色的沙发和整个空间显得相得益彰，让整个空间更加和谐。

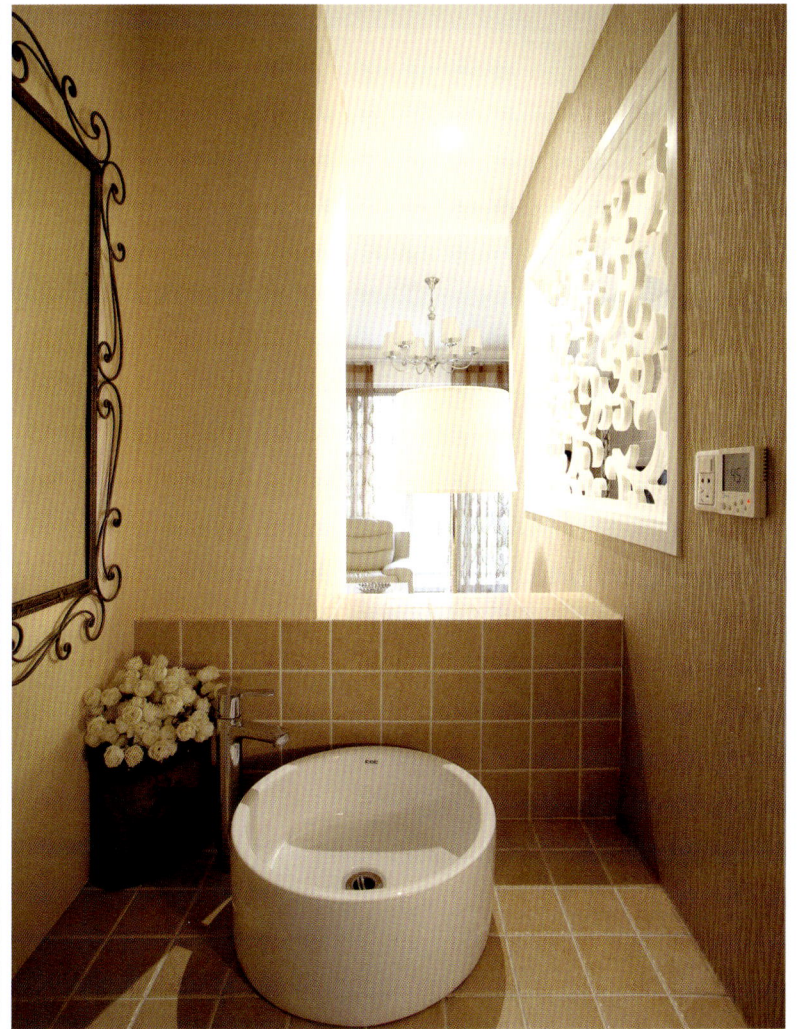

Guizhou Bailing Fashionable World, No.3 Fine Decoration Sample Apartment

贵州百灵时尚天地精装公寓样板房三

项目地点：贵州
项目面积：57.83平方米
设计师：曾涛
设计公司：贵州峰上室内外设计工程有限公司
主要材料：马可波罗砖、罗马利奥砖、墙纸、西奈珍珠石材、石英石、科勒洁具

The design of the apartment for single is full of fluctuant and smooth curved and round lines, and through the careful design of space, materials and colors, the designer creates a graceful and romantic atmosphere of living space.

The living space plays role of multiple functions in one set, including receiving room, dining room, bar, wash room, and other compound functional spaces, so the small space features complex and changeable functions through the design. In the bedroom, a big unique circular bed is introduced, and the ground is fully paved with deep color dense carpet to define spaces. The ceiling in the bedroom is concave round, coupled with the overall round lines in the space, and a golden droplight is introduced to further heighten the importance of bedroom and the luxurious atmosphere.

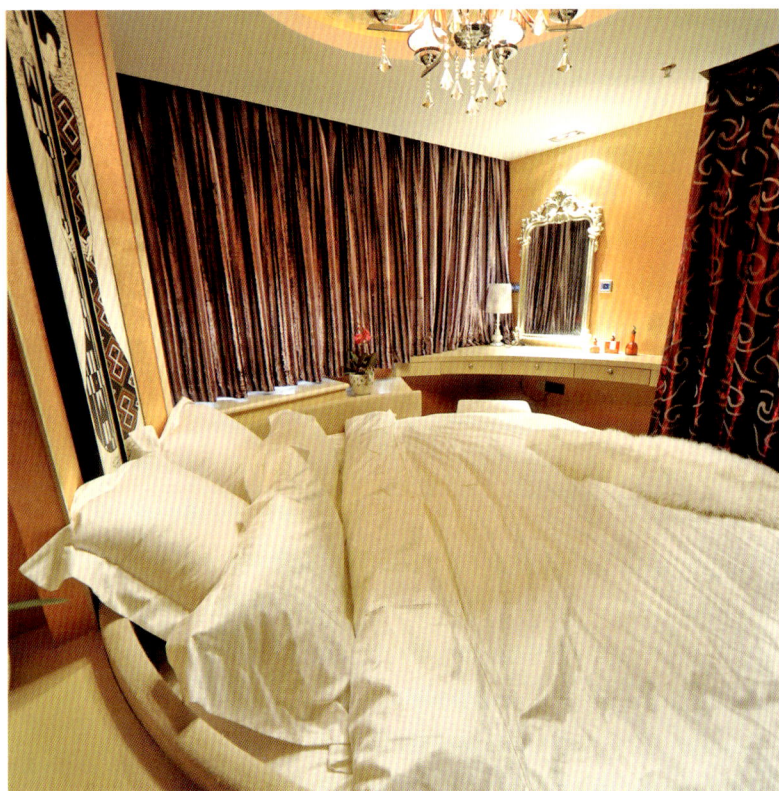

这套单身公寓的设计中，充满了起伏而流畅的弧形、圆形线条，通过对空间、材质、色彩等方面的精心设计，营造了居室婉约的浪漫氛围。

起居空间集多元功能于一体，包括会客厅、起居室、酒吧、盥洗室等多种复合功能空间，使面积并不大的空间通过设计实现复杂而多变的功能性。卧室采用了独特的圆形大床，地面采用了深色而茂密的地毯满铺形式界定空间。卧室天花做出了圆形凹进，与室内整体圆形线条相协调，更采用了金色的吊灯形式，进一步烘托卧室的重要地位与奢华氛围。

The Pure Space

纯境空间

项目面积：97平方米
设计师：陈龙
设计公司：南京火龙空间设计工作室
主要材料：橡木、罗丹瓷砖、茶镜、钢化玻璃、志邦橱柜、
生活家地板、乐家洁具等

The designer first made considerable adjustments to the original structure. The two seemingly separate space, the kitchen room and the studio, are combined into an integrated space by the decorated door pockets and glass flora patterned walls. As a result, the size of the kitchen is enlarged, and the limited space is fully used.

On the basis of the original black, white and gray places, dark coffee colors are added into it, and as a consequence, the cool and sober living space is infused with a sense of fashion, manifesting the touch of modernity and warmth. This design has made flexible adoption of glass and mirrors to extend the effect of lighting and spatial third dimension. The special form of lights, the mirror surfaced light box with oval and rectangular shaped etched patterns, the mosaic arranged lines formed by black and white interphase enhance the dialogue of spatial atmosphere.

Many details give focus to the conciseness and grandness so that the precipitation and visualization of the design are well revealed.

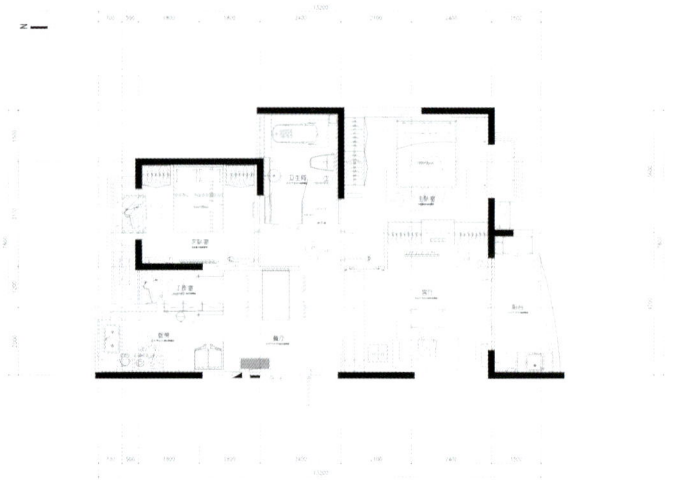

设计首先对结构做了较大的调整，厨房与工作室两个看似独立的区域，被装饰门套及玻璃花卉图案墙组合成一个整体，增大了卫生间的面积，充分地利用了有限的空间。

设计在原有黑白灰色彩面域基础上加入深咖啡色，为素雅的生活空间注入一股时尚气息，彰显了该设计现代、温情的一面。设计对玻璃和镜面的灵活运用，使采光效果和空间立体感得到延伸。独特的灯具造型、椭圆与矩形蚀刻图案的镜面灯箱、黑白相间的方砖马赛克线条，渲染了空间之间的对话。

众多细节突出了整体设计的简洁与大气，体现了设计的积淀与构想。

The Oriental City

东方名城

项目面积：120平方米
设计师：叶强
设计公司：宽北设计机构
主要材料：i-box家居连锁、进口壁纸、仿古砖、天然大理石、橡木修色

This design restructures the space with the modern classic oil paining style, uses a mixture of texture elements, such as wood, leather and fabric, and combines the both close and loose grains of wood, the soft and smooth textures of leather, and the fit and complex characteristics of fabrics, manifesting the sober and grand character of the interior space. In terms of colors, the designer adopts the warm and sedate colors, such as orange, beige and brown, to form a harmony with the white sofa and other furniture, creating a quite individualized household atmosphere. The crystal drop light in the living room becomes the focal center and key element that lead your sights, and harmonizes well with other elements. The silver and white craft furnishings match perfectly with the classic atmosphere of the interior space.

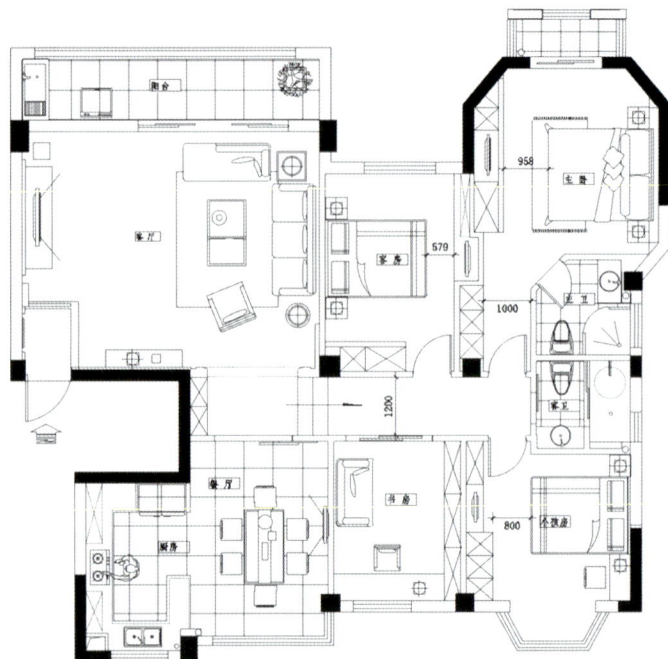

　　该设计以现代经典油画风格重建空间，设计中综合使用了木料、皮质、织物等肌理元素，结合了木料疏阔有致的纹路、皮质细腻柔软的质感、织物熨帖错综的特性，彰显了室内空间沉稳大气的性格。在色彩方面，设计师采用了橘色、米色、褐色等温暖而沉稳的色调，与白色的沙发等家具相协调，创造了一种充满个性的家居氛围。客厅的水晶吊灯作为视觉的中心点及统领视线的关键元素，与其他元素和谐搭配。银色、白色的工艺摆饰与室内的古典氛围融合得恰到好处。

Guizhou Bailing Fashionable World, No.2 Fine Decoration Sample Apartment

贵州百灵时尚天地精装公寓样板房二

项目地点：贵州
项目面积：57.83平方米
设计师：曾涛
设计公司：贵州峰上室内外设计工程有限公司
主要材料：马可波罗砖、罗马利奥砖、墙纸、西奈珍珠石材、石英石、科勒洁具

The project is a small size apartment for single, and the designer focuses on the good treatment and care on the liver's soul, where the elegant and simple lines are introduced in many places, coupled with elegant floral patters and warm and thick wood texture.

The interior colors are bright and warm mainly with warm beige and yellow, and in order to neutralize the interior bright colors, deep brown and grass green in the same color system are introduced to make the interior colors more natural and harmonious. To meet the luxurious need of apartment for single, the designer introduces droplight with simple lines in the living room to guide the space, and take care of conveying a sense of luxury in the treatment of headboard wall and selection of fabric on curtains.

该项目为小户型的单身公寓，设计师着重对居者心灵的善待与呵护，在设计中多处使用了优雅、简洁的线条，搭配雅致的花卉纹样与温厚的实木质感。

　　室内色彩明亮而温馨，多处使用了温暖的米色、黄色，为了中和室内明亮的色彩，更是搭配了同色系的深棕色、草绿色，使室内色彩更为自然和谐。为了兼顾单身公寓的豪华需求，在起居室，设计师选择了线条并不繁复的吊灯形式，统领起居空间，并在床头墙面的处理上、窗帘纱帐等布艺的选择上，兼顾了对奢华感的传达。

Portofino Apartment

波托菲诺住宅

项目地点：墨西哥
项目面积：440平方米
设计师：Javier Sánches、Pola Zagga、Larissa Kadner、Brenda Ochoa
设计公司：JS[a]
摄影师：Jair Navarrete、Pim Schalkwijk

Located in a major residential development within Bosques de las Lomas in Mexico City, this apartment was bought by a family of three, while it was in the early stages of construction. All bathroom and kitchen services, however, had to remain in their original place. The challenge here was to keep the space as clean and as open as possible, in order to take advantage of the spaciousness of the rooms and to bring out the space where light comes in along the perimeter of the apartment. The designers also wanted to differentiate between the private and the common areas by playing with contrasts in the finishes.

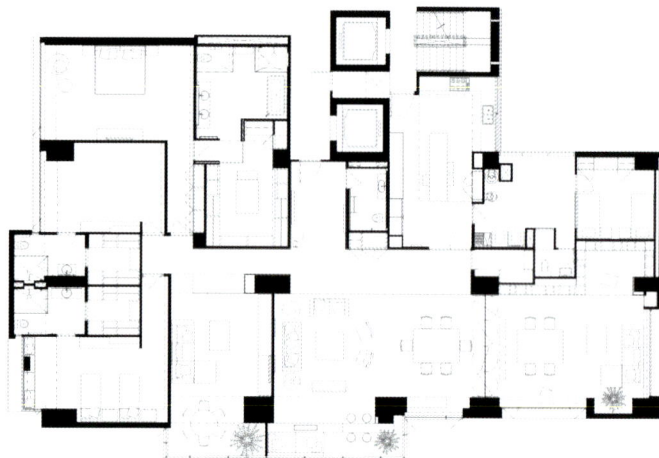

0   5m

　　本案位于墨西哥城拉斯洛马斯的主要居住新区，一个三口之家买下了这幢房子。当时房子处于建设初期阶段，所有卫生间和厨房设施不得不保留其原来的位置。设计师面对着一个挑战，要在保持空间的清洁和开放的前提下，充分利用房屋的宽敞度，并在光线沿着建筑的四周照射进来的地方展示室内空间。设计师还希望通过漆面的对比来区分私人空间和公共区域。

"White" Romance

"白"色浪漫

项目地点：北京
项目面积：150平方米
设计师：吴巍
设计公司：北京东易日盛装饰股份有限公司
主要材料：帕格尼尼白橱柜、Haro地板、HA IRIS azuv瓷砖、意德法家壁纸

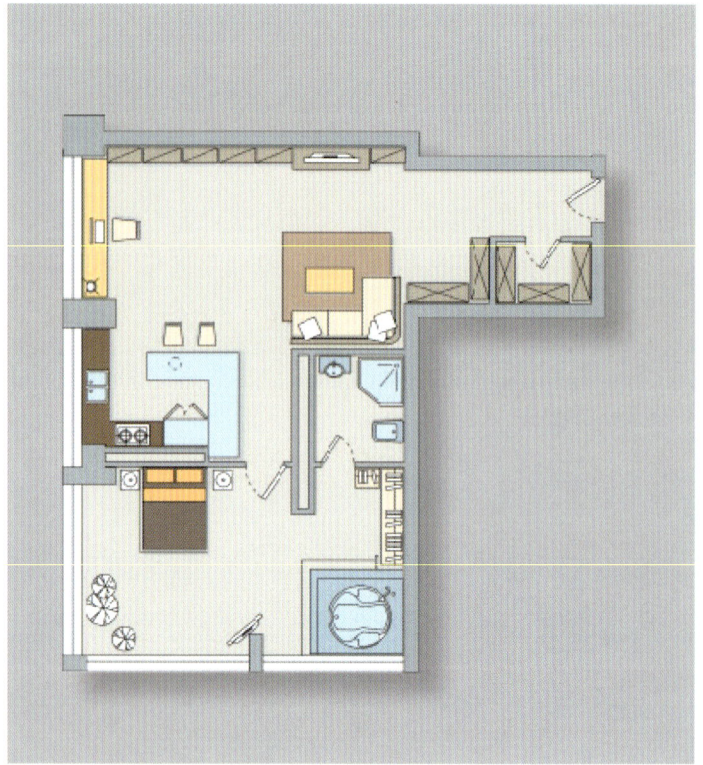

The whole space is designed to be modern and simple style in white tone, and the simple lines are used to create modern and stylish atmosphere in the space, providing a feeling of relaxation.

The entrance hall, living room, study and kitchen are designed to be open, divided by furniture and partial platform. In the living room, the pure white storage cabinet is integrated with the wall into a whole, and the TV background wall is designed with hidden storage lattice, playing a good role of decorating. The kitchen is located just in the middle of the living room, expanding from the center in the form of island. The bath is separated from the master wash room, put in a corner of the master bedroom, and the half-underground round bath with white stone countertop is coupled with droplight and curtains, cleverly integrated into the master bedroom.

　　设计师将整个空间定位成现代、简约风格，以白色为主调，运用简洁的线条营造现代、时尚的空间氛围，给人轻松的感觉。

　　门厅、客厅、书房、厨房在空间上都处理成开放的形式，利用家具及局部的地台在空间上做了划分。客厅纯白色的收纳柜与墙面融为一体，电视背景墙被设计成了隐蔽的收纳格，也起到了很好的装饰作用。厨房位于客厅的正中间，以岛台形式居中展开。设计师将浴缸从主卫中独立出来，放到了主卧的一角。下沉式的圆形浴缸、白色的石材台面配以吊灯和窗帘，巧妙地融入了主卧空间。

Guizhou Bailing Fashionable World, No.4 Fine Decoration Sample Apartment

贵州百灵时尚天地精装公寓样板房四

项目地点：贵州
项目面积：108.23平方米
设计师：曾涛
设计公司：贵州峰上室内外设计工程有限公司
主要材料：马可波罗砖、罗马利奥砖、墙纸、西奈珍珠石材、石英石、科勒洁具

Through careful handling and management, the designer weakens the polygonal space, and further enhances the interior flow lines, making the space more reasonable and practical.

The interior design is neo-classical style, with strong deep colors as the main colors, interpreting the magnificent character of neo-classical. The furniture is high-profile and elegant with calm temperament, where the purple velvet sofa texture features subtle changes on vision with the changes of angle and light, and the octagonal tea table is decorated with silver texture to form the visual center of the space, creating a multidirectional atmosphere of the living space. The natural patterns on the sofa background wall and walls in the bedroom are ornate and delicate, filling the apartment with a unified romantic atmosphere.

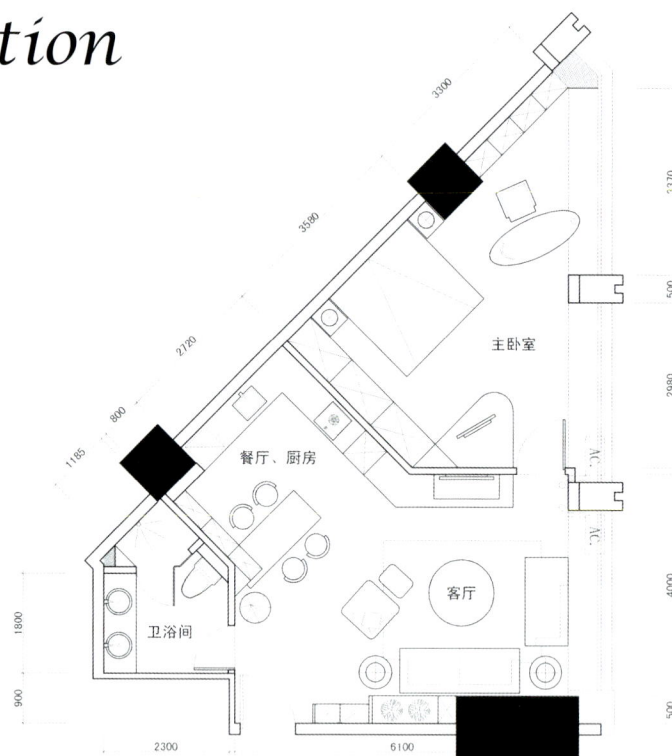

通过精心的处理与组织，设计师通过设计弱化了这个多边形空间，将室内的流线进一步强化，使之更为合理和实用。

室内设计采用了新古典主义风格，以浓重的暗色为主色调，诠释了新古典主义的华丽特质。家具高调而优雅，气质泰然。紫色的丝绒沙发纹理随着角度、光线的变化在视觉上产生着微妙的变幻，八边形的茶几则采用了银质质感，形成了该空间的视觉中心，更营造了起居空间的多向围合氛围。沙发背景墙、卧室墙面等处的自然纹理的图案华丽而精致，更使整套公寓弥漫着统一的浪漫气息。

Humanity

人·文

项目地点：台北
设计师：俞佳宏
设计公司：尚艺室内设计有限公司
摄影师：徐凯威

The designer avoids using too extensive piling up and creating design feeling concept in his design. With this project, the designer internalizes the design concept, so that the space is upgraded to the intangible human quality and connotative layer. In the light Zan style created deliberately by the designer, the wood louver door chooses a kind of delicate lines that echo the modern humane culture. In such a space, the designer also combines the fog-surfaced wood-stone, and with the two natural materials of stone and wood, the designer creates a cozy and easy residential atmosphere.

Through the management of content materials and the creation of interior atmosphere, the designer makes this design not only a simple gathering of pure materials, but also a humane art created by care.

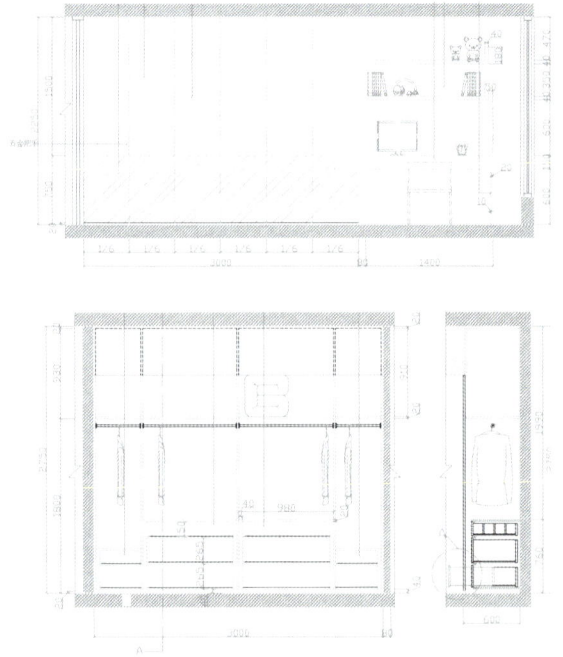

设计师在设计中避免了过度堆砌和营造设计感的概念，通过此作品将设计思路进行内化，从而使空间提升到无形的人文质感与涵养层面。在设计师精心营造的淡淡禅风里，木百叶门片采用了一种细腻的线条，呼应出现代的人文情怀。在这样的空间中融合了雾面木化石，通过石与木两种自然素材营造惬意与轻松的家居氛围。

设计师通过对内容元素的拿捏与室内氛围的营造，使设计不再是单纯素材的简单拼凑，而是用一种关怀营造出"家"的人文艺术。

Ye's Residence of Taoyuar

桃园叶公馆

项目面积：135平方米
设计师：黄建华、黄建伟
设计公司：黄巢设计工务店
主要材料：银狐大理石、绿建材木皮板、清玻璃、烤漆玻
璃、墨镜、超耐磨地板、系统柜、烤漆、夜光壁纸

In the selection of colors, the designer chooses the collocating strategy mainly with gray and white colors, and the whole space is decorated with gray veneer and dyed-white veneer to keep unified and harmonious. The simplification of interior colors and lines creates the modern and simple sense of the space.

The TV wall is designed with large area of penetrating type to create a luxurious atmosphere, and other corners can be apprehended at a glance from the public space. The piano room is designed with large-capacity bookcases to show the owner's collection of books and make the space multi-functional. The TV main wall is decorated with gray veneer coupled with silver-fox marble of white bottom, and the design of penetrating grille allows each space interconnect with each other. The partition between the kitchen and dining room eliminates the sense of pressure in the space, and the embedded large-area clear glass takes the roles of lighting and obstructing cooling fumes into account.

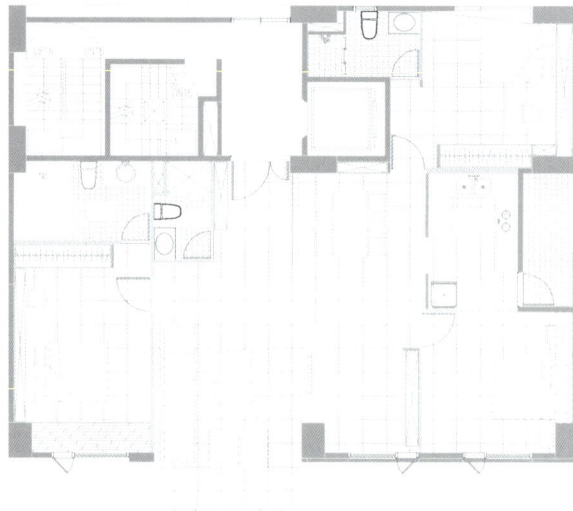

在色彩的选择上，设计师采用了以灰白两种颜色为主的搭配方案，整个空间使用灰色木皮及染白木皮做搭配，统一而和谐。室内颜色、线条的单纯化营造了现代简约的空间感。

设计采用大片穿透式的电视墙设计，营造豪宅氛围，在公共空间各处皆可对其他角落一目了然。琴房内设计了大容量的书柜，展示屋主的藏书，使空间多功能化。电视主墙灰色木皮搭配了白色底的银狐大理石，穿透式的格栅设计让每个空间得以相互连结。厨房及餐厅隔墙的处理消除了空间的压迫感，嵌入式的大片清玻璃兼顾采光及阻隔油烟的作用。

Hankyu Rokko Residence

阪急岛公寓

项目地点：神户
设计师：Elena Galli Giallini
设计公司：ELENA Design Office

The theme for this project was the design of "order made" apartment units for a multifamily residential complex near Kobe, Japan.

The design approach centers on a high level of customization to meet the investor's objectives. The program required an attentive research on Japanese customer's current and future needs, trends of life style and market orientation. The goal was to propose a different image of the house, in Japan, for a modern life style, with focus on user-friendliness, functionality and originality.

The layouts are rational and efficient, privileging performance and solutions suitable for a more efficient, modern and flexible use of the house. Plan versatility allows corresponding to the needs of different household types. Adaptability to the changes over time and to the evolution of family requirements and composition were key concerns to the project. Furniture, fittings, fixtures up to the smallest details, all were specifically designed and custom made for the apartments of this complex. Natural materials and delicate, light colors were chosen to feature an environment with delicate tonality, understated refinement and a sense of naturalness. Just few accent points in vivid, warm colors and brilliant finishes strengthen and characterize the whole. Neat design, sharp details define the essentiality of the ambiance.

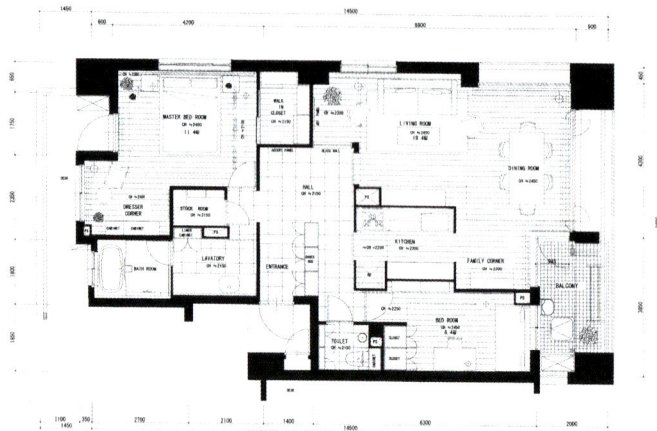

本案以"定制"设计为主题，在日本神户附近为多户型家庭住宅营造复式公寓。

设计方法强调了高层次定制，以实现投资者的目标。根据本案的要求，设计师特别对日本客户当前和未来的需求、生活方式趋向及市场导向做了调查研究，其目标是在日本为现代生活风格提供不同的住宅形象，强调用户友好性、功能性及独创性。

本案布局合理有效，特殊的性能和处理方法更适应了住宅的有效、现代、灵活的使用要求。多功能计划满足了不同家庭类型的需要。对时间变化及家庭需求和组成演变的适应性是本案要解决的关键问题。家具、装饰、装置及微小的细节，为复式公寓专门设计和定制。设计师选择了天然材料和细腻的浅色系来突出色调精美、低调完善、自然感的环境特点，生动温和的色调和明亮的设计中些许重点加强并定性了整体空间。简洁的设计、清晰的细节定义了空间气氛的本质。

Black and White

黑白

项目地点：福州
项目面积：150平方米
设计师：蒋兴达
设计公司：福州吉祥如意装饰工程有限公司
主要材料：L&D陶瓷、大自然地板

In the selection of materials, the designer takes low-key strategy, and creates a restrained texture by the sharp contrast between black and white to set off the sense of design in the space. The design of the whole space is neat and bright without extra lines, and the contrast relationship between black and white depicts the fashionable vocabulary of the space, with less decorative lines, but more depth and intellectual temperament.

The design of the study room is warm and elegant, and the solid wood flooring gives people a warm and comfortable feeling. The neat and practical storage cabinet can be used to display books and crafts. The open-style kitchen leads to the dining table, so the small kitchen is not suppressed anymore, and the space is spacious and bright, creating a cozy and relaxing elegant atmosphere in the space.

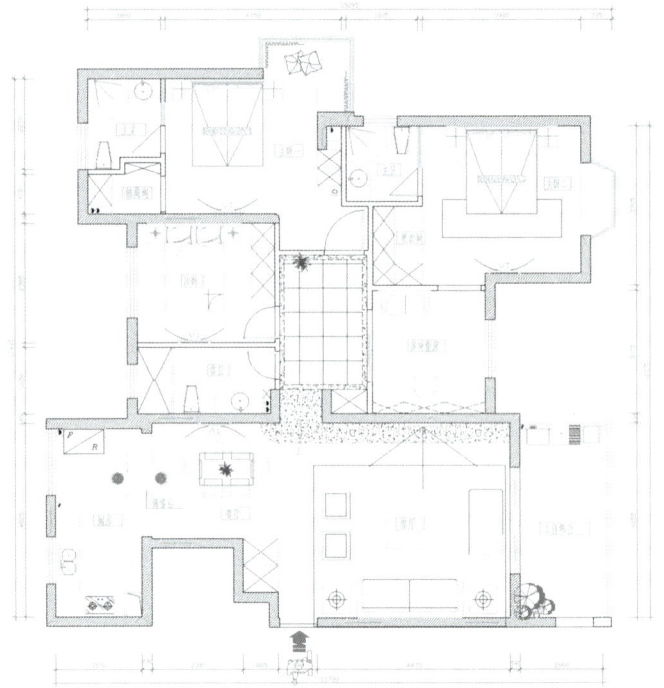

　　本案材质选择方面采取低调策略，藉由黑白强烈的对比关系，构建出内敛的质感，衬托空间的设计感。整个空间的设计简洁明快，没有多余的线条。黑白色交错的对比关系，刻画出空间里的时尚语汇，少了装饰线条，却多了深度与知性气质。

　　书房的设计温馨而优雅，实木地板给人温暖舒适的感觉。收纳柜的设计简洁实用，可供书籍和工艺品陈列。开放式的厨房通向餐桌，使狭小的厨房不再压抑，空间宽敞而明快，营造出空间悠闲而惬意的优雅氛围。

Neo-classical Style

新古典风格

项目地点：上海
设计师：宋建文
设计公司：上海设计年代
主要材料：抛光砖、木地板、水晶灯

This is a neo-classical style house. The designer fixes the interior furniture mainly with European style, supplemented by classical and modern styles on details, and the organic integration of many styles forms this neo-classical living space.

In the living room the furniture is mainly gold and white, and the collocation of the two colors makes the living room magnificent, full of noble atmosphere. The crystal droplights in the dining room and living room are in common style, strengthening the classic and luxurious theme together. The design of the master bedroom is full of noble and elegant taste, where the purple window screen is coupled with bedding of the same color, and elegant and chic bedside lamps, and you can feel the romantic flavor of European style everywhere.

这是一套新古典风格的居室。设计师将室内的家具风格定位为以欧式风格为主，在细节之处又辅以古典风格和现代风格，众多风格的有机融合，形成了这方新古典居室空间。

客厅家具的颜色以金色和白色为主，这两种颜色的搭配使客厅金碧辉煌，充满了高贵的气息。餐厅与客厅的水晶吊灯风格一致，共同强化了古典奢华的主题。主人房的设计充满高贵优雅的味道，紫色窗纱搭配相同色系紫色的床铺及优雅而别致的床头灯，处处都使人感受到欧式风格的浪漫情怀。

Yintai Center

银泰中心

项目地点：北京
项目面积：198平方米
设计师：王奇
设计公司：尚层装饰（北京）有限公司
主要材料：瓷砖、乳胶漆、大理石、玻璃

The project is based on a simple style, and the designer seeks a beauty of balance in the vertical and horizontal design experience. The elaborate technology and sophisticated materials are used to show the unique refinement and personality of industrial society. The introduction of glass products and metal leather make the contemporary space transparent, fully showing the elegant taste without any extra refined style, and the leather sofa of neat style features strong texture. The transparent glass tables are used repeatedly to keep united on vision, and the simple lines and collocation of colors provide various scenes for the home living. The bedroom is in semi-open pattern. The beddings are decorated with black and white striped patterns, settling the noisy of life, and the careful handling of light softens the strong visual differences of black and white colors.

本案采用简约的风格基调，在横平竖直的干练中寻求着一种平衡的美感。用精细的工艺和考究的材质展现出工业化社会中独有的精致和个性。玻璃制品和金属皮革的使用，让富于时代感的空间内外通透，在没有任何多余的精炼风格中，高雅品位展露无疑。造型干净利落的皮质沙发，质感极强。重复使用的透明玻璃桌为空间带来视觉上的统一，简洁的线条和配色，为家居生活提供了丰富的场景。卧室采用了半开放式，床品也选择了黑白条纹图案。沉淀了生活中的喧器。灯光的精心处理柔化了黑白两色强烈的视觉差异。

The Quiet Residence of Stone, Wood and Water

石木水静宅

项目地点：台北
设计师：林琮然
设计公司：阔合国际有限公司

The designer adopts both the hidden and open styles to construct a connotative spatial order. The bland doors craftily hide the private space and create a broad space, and also make the whole layout full of adjustable flexibility. The slope folding door crosses on the conjunction of the inside and outside, so that it blurs the relationship of inside and outside and produces a boundary and extension that are hard to distinguish. Such changes increase the multiple functions of the usage area. The reverse side of the black wall at the entrance foyer is the bright wall that extends to the outside of the balcony. The choice of two sides of one wall applies the extension of different materials not only widens the vision but also introduces nature into the interior space. The thought of breaking the pattern makes the family life throughout each corner of the space. The integral cabinet can provide big storage space in the master bedroom, using natural patterned materials in a large size to create the broadness of bearing of the owner. The integral environment takes on the unconventional quality of a bright and pure residential space, and adds much happiness to the peaceful nature.

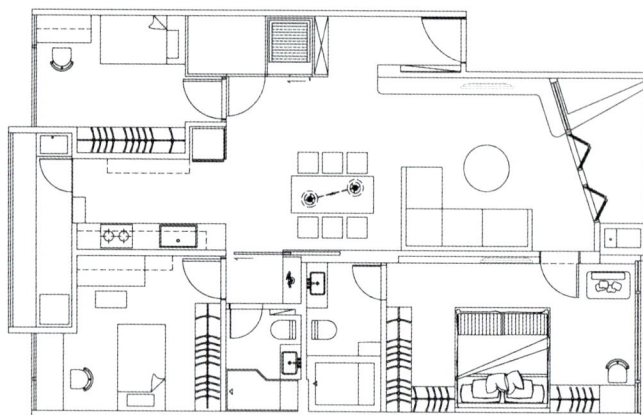

设计师运用了隐藏与开放的手法构建出隐含的空间秩序。暗门巧妙地将私密空间隐藏起来，创造出更宽广的空间，也让整体布局充满了可调整的弹性。客厅与阳台间的斜面折叠门横跨在内与外的交界处，模糊了里外的关系，产生难以辨别的界线与延伸。这样的转变增加了使用面积的多重运用方式。入口玄关黑色墙面的背面是延伸至阳台屋外的亮色墙壁，一体两面的墙壁运用相同材质的延伸，不仅开阔了视野也将大自然引入室内。打破格局的空间思考将家庭生活延伸至每个角落。主人卧室内整体储物柜提供大量的储藏空间，利用大面积自然材质纹路打造屋主的宽广气度。整体环境展现居家空间明亮与洁净的脱俗气质，在平和自然之外增添欢乐气息。

Maple Forest Residence

枫林丽舍

项目地点：重庆
项目面积：130平方米
设计师：王品
设计公司：重庆翰艺室内装饰设计工作室
主要材料：瓷砖、天然石、实木板、乳胶漆

This project takes the Mediterranean style as the key tone, and employs many Mediterranean elements like the arched doors. The elements such as the solid wood walls with light colored and vertical stripes, the cream-colored cloth sofa and the bright blue cushions bring a fresh ocean aura to the interior space.

On the basis of this style, the design employs a mixed style appropriately and chooses bright colors and exquisite details to reveal the high quality of life attitude. In order not to make the mixed style be disordered, this design attaches importance to the visual balance, and makes the furniture and ornaments of diverse styles become an integrated one on the combination of the whole. In terms of the floor, it adopts the mosaic style with more materials and approaches, which plays a good role in segmenting the areas.

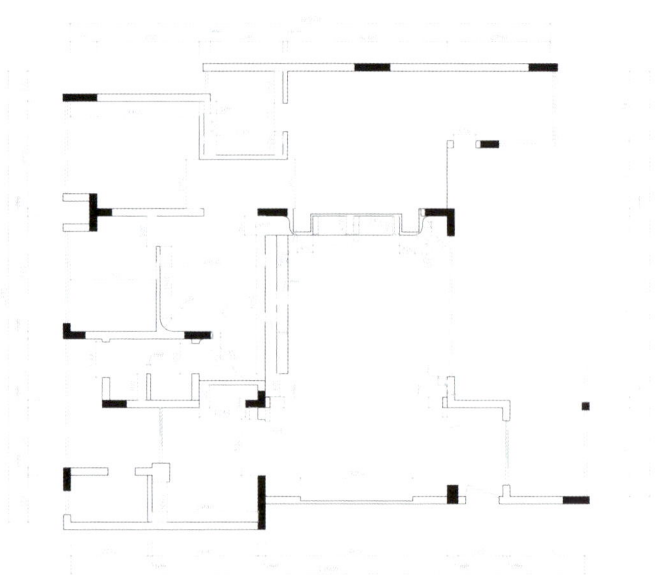

本设计以地中海风格为基调，运用了拱形门洞等多种地中海风格元素。浅色纵向条纹的实木板墙面、米色布艺沙发、明亮的蓝色抱枕等元素，给室内带来了一股清新的海洋气息。

在此风格基础上，设计适当运用了混搭的手法，用明快的色彩及精致的细节体现高品质的生活态度。为了使混搭风格不显得混乱，设计注重视觉的对称，使风格多样的家具和饰品在总体组合形式上形成了统一。地面采用马赛克搭配多种材质及不同的拼贴方式，起到了很好的区域划分作用。

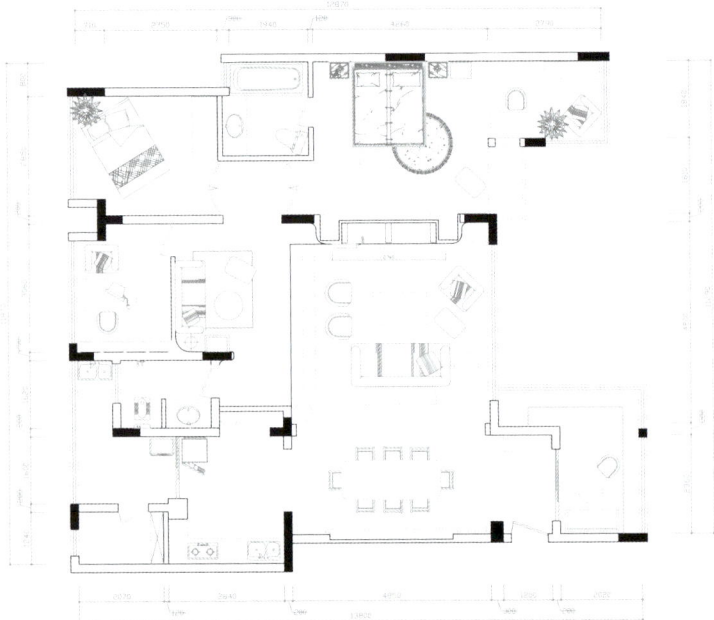
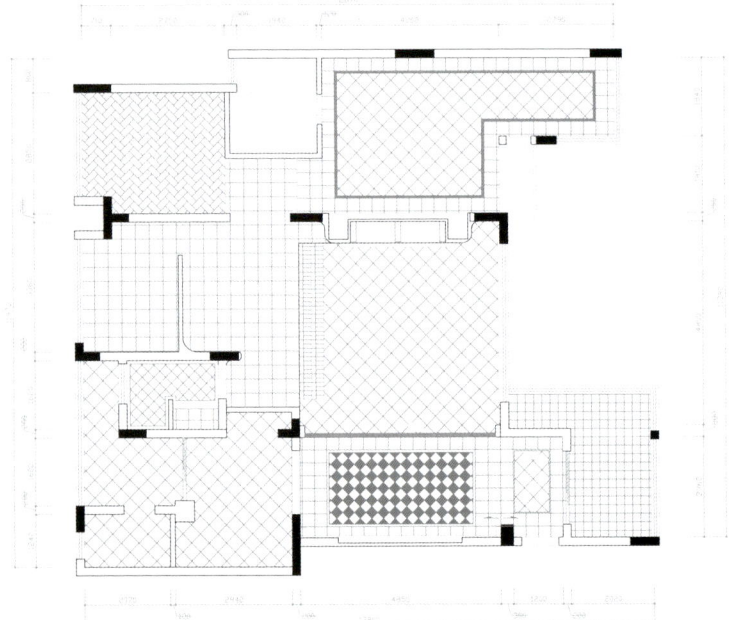

Liang Xian Residence

亮贤居

项目地点：香港
项目面积：约108平方米
设计师：郑勇威
设计公司：爱家设计有限公司
主要材料：直纹橡木、真皮、西班牙米黄及啡网云石、灰镜、墙纸

This project takes the modern classic design style as the leading style, and the whole project chooses wood flooring and wood colored furniture to create an easy and cozy living experience.

The crystal chandelier in the dining room forms a progressing impression that exists between deficiency and excess, which contrasts finely with each other, and makes one feel, at the same time, the fashionable and pleasant atmosphere. The black moulded glass creates a rich organoleptic experience and an easy and comfortable dining atmosphere, which sets the position for the distinct style of the interior space, and through the connotative perceptual information, clearly interprets the extravagant cross-domain concept. The extending design of the private space and the planning of the hidden collecting function reduce the suppressed constriction caused by excessive cabinets. The fitting room takes the wall ornaments with classical connotations as the main impression, which appeals romantic and cozy.

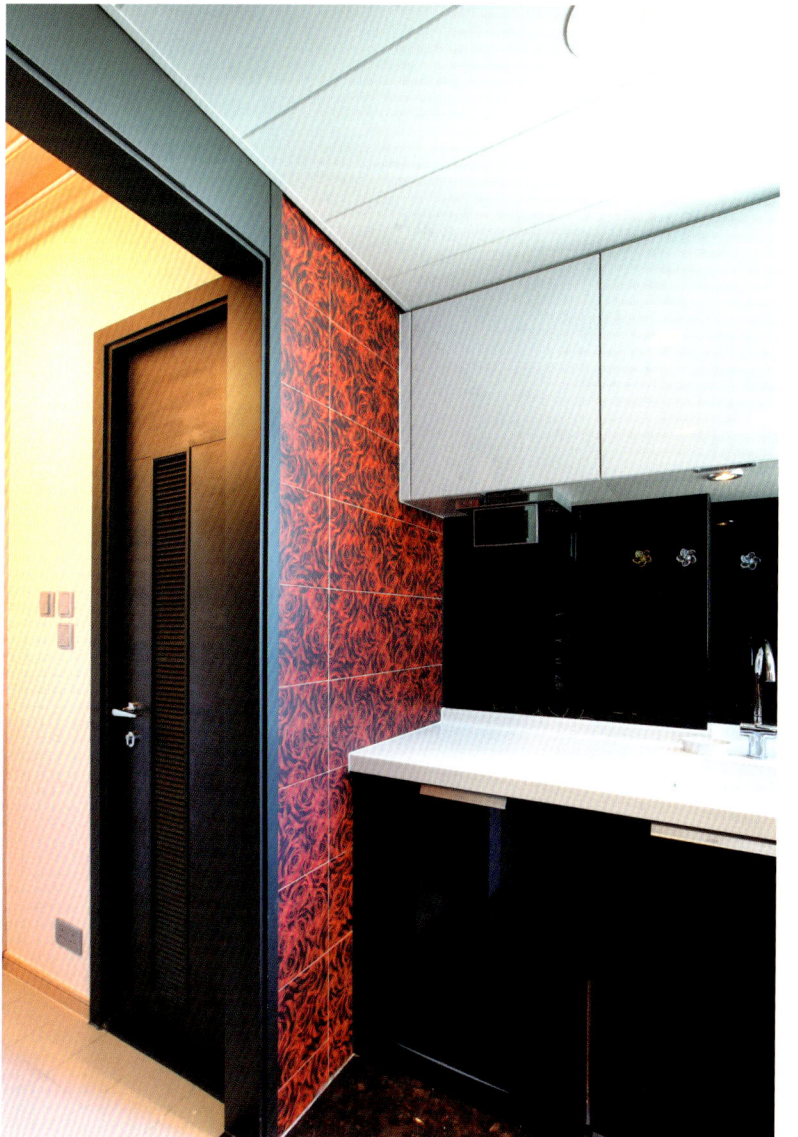

该设计以现代经典设计风格为主导，整体采用木地板及木色家具，创造出轻松、温馨的家居体验。

餐厅的水晶吊灯形成虚实之间的渐进表情，相映成趣，同时令人感受到时尚及欢愉的气氛。黑色造型玻璃创造出丰富的感官体验，营造出一种轻松舒适的就餐氛围，为室内鲜明的风格进行了定位，透过隐含的感性信息，清楚地解析出跨域的奢华思维。私密空间的延伸设计、隐匿的收纳机能规划，减缓了过多的柜体对于空间造成的压迫感。更衣间以具有古典韵味的墙面装饰作为主要表情，浪漫而温馨。

Western Maple Egret Residence, Fuzhou

福州丹枫白鹭

项目地点：福州
项目面积：145平方米
设计师：施继诚
主要材料：仿古砖、硅藻泥、松木、马赛克等
摄影师：施凯

The lifestyle that the occupant couple expects is to embrace nature just like living in a wood cottage in the forest, enjoying the pastoral life.

After setting up the design direction, the designer began his deliberate planning for the layout, modeling, colors and furniture placement in the interior space. The layout is transparent and bright, and the moving lines are smooth. As requested by the owners, the designer sets a multi-functional area in the master bedroom, which provides plenty room for either working and study, or leisure and recreation. In terms of the modeling, in order to overcome the lack of height of the ordinary apartment, the designer has tried a few non-mainstream reduction attempts for the strict European style, and makes this project more amiable and acceptable, and the whole style more unite. The living room is in yellow and red color tones, which is bright and cozy. In the bedroom, however, the elegant and romantic light blue, along with the refreshing and invigorating light green, produces the tranquility and peacefulness particularly owned by oceans and forests. The carefully selected furniture and furnishings enhance the overall effect, and adorn the pastoral atmosphere of this apartment.

业主夫妇期待的生活方式是如同居住在森林木质小屋里一般，可以拥抱自然、享受田园生活。

在明确了设计方向后，设计师在室内的布局、造型、色调、家具陈设等方面进行推敲，使布局上通透明亮、动线顺畅。设计按照业主的要求在主卧室安排了一块多功能区域，不论用于工作学习还是休闲娱乐都能游刃有余。造型上为克服普通公寓层高不足的缺欠，针对严谨的欧式造型元素做了一些非主流的弱化尝试，使其更加亲切、易于接受，使整体风格更加统一。客厅使用黄红色调，明快温馨，卧房里优雅浪漫的淡蓝与清新爽快的浅绿带出大海与森林特有的宁谧感受。搭配精心挑选的家具与陈设为整体效果增色，装点出田园氛围。

Happiness & Environment

乐·境

项目地点：南京
项目面积：135平方米
设计师：董龙
设计公司：DOLONG董龙设计
主要材料：拿铁家具、包豪斯餐桌椅、硅藻泥、进口墙纸、进口
地板、灰镜、杜拉维特洁具等

Minimalism, however, does not mean simplicity. In such a minimalist space, white color is chosen as the key tone. The designer takes advantage of the white doors, white flooring tiles and white walls to reach a good harmony. On the basis of white, it is blended with similar colors like the cream color. The design adds longitudinal striped styled gray mirrors to decorate the walls to enhance the sober and sophisticated atmosphere of the space. As a consequence, the whole project is elegant and clean as well. Some metal-textured hanging ornaments are decorated on the walls, which set off the elegance of the space. The tawny red sofa is more outstanding in such a pure colored space. As an important element of the functional space and the visual center of the space, the use of such color elements is just to the point.

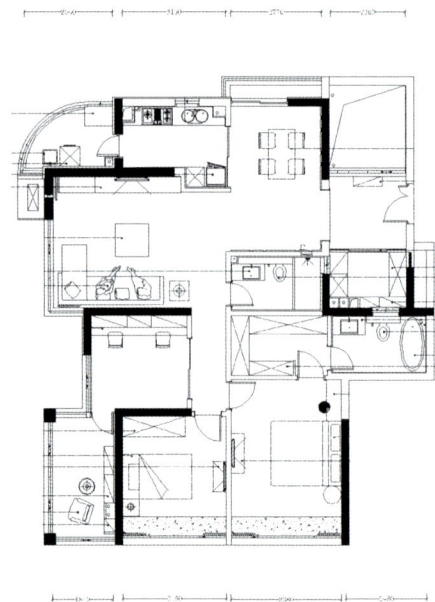

简约，却不意味着简单。这个简约的空间，以白色调为主。设计师利用了白色的门、白色地砖与白色墙面营造良好的协调性。在白色的基础上融入米色等相似色彩。设计添加纵向的条形灰镜点缀墙面，增添了空间沉静的氛围，很素雅，也很干净。墙壁上点缀了一些具有金属质感的挂饰，烘托了空间的高雅之感。茶红色沙发，在这一方素色的空间中格外引人注目，作为功能空间的重要元素以及空间中的视觉中心，此种色彩元素的使用恰到好处。

Jiang NanWater City Island, Fuzhou

福州江南水都丽岛

项目地点：福州
项目面积：120平方米
设计师：卓新谛、于斐
设计公司：福州合诚环境艺术有限公司
主要材料：大理石地砖、墙纸、黑镜

The designers of this project adopt the modern design concept and have made an in-depth treatment to the space. While extending the modern life space, at the same time, they have combined a classical nobleness into the project.

The whole space uses black, white and gray colors as the key tone. The walls pieced with marble tiles and striped stones match well with the light colored flooring. The light colors of the space are so harmonious and show a rich texture of different layers, and further form a sharp contrast with the black colors of walls and furniture. The colors of the space supplemented with coffee and silver colors make the space combine with many characteristics of both modernity and reminiscence. What is more special is the bright green that boldly used by the designers in the stairway area, which makes the whole space filled with an irresistible vigor.

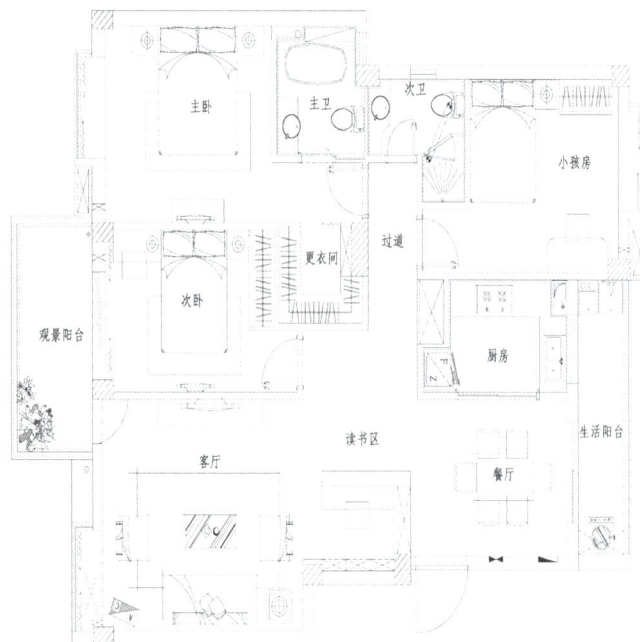

本案的设计师利用了现代设计的理念，对空间进行了深入的处理。在对现代生活空间进行延续的同时，融入了古典贵族气质。

整个空间以黑白灰为基调，大理石地砖、条形石材拼合的墙面搭配了浅色木地板，使空间中的浅色部分互相协调，呈现了不同层次的丰富质感，更与墙面、家具的黑色部分形成鲜明对比。空间色彩辅以咖啡色及银色，使空间融合了现代和怀旧特质。尤其是设计师在楼梯区大胆地使用了鲜艳的绿色，让整个空间融入了一股无法抵挡的活力。

Yixing Jing He Family

宜兴景和人家

项目地点：江苏宜兴
项目面积：140平方米
设计师：宋春吉
设计公司：常熟吉恩设计事务所
主要材料：波浪板、饰面板、木纹砖、大理石、磨花镜子
摄影师：文宗博

The designer applies the idea of "less is more" to the whole design, which takes white as the key tone aided with black to create a bright and neat space visual effect.

In this project, the complicate overuse of shapes and materials is abandoned. In terms of decoration, the simplified design concept is employed to create the format of the space, with which design elements of coziness, comfort and understatement are blended. While in terms of ornament, the designer chooses the design technique of addition to render the class of life, which fully highlights the disposition of the black-and-white style' sophisticated, comfortable, natural and bright. The pure white waved plate is just like the wavelet of the lake outside the window, with which the designer leads the outdoor scene into the interior space in an abstract way to enhance the flexibility of the space. It is nothing but "The mountain is more secluded while the birds are singing", and this brings out the best in each other with the theme of this design — waves.

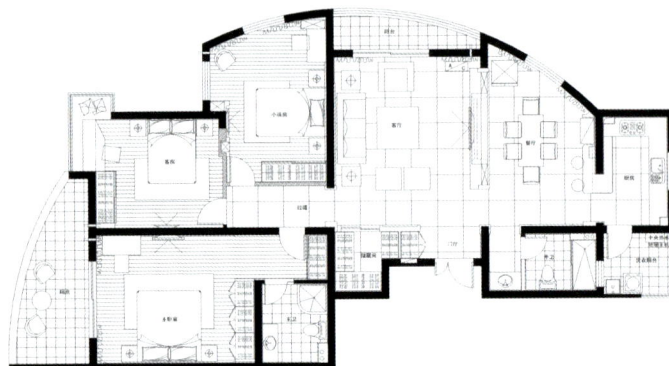

设计师将"少就是多"的理念应用到整体的设计中，以白色为主基调，辅以黑色，营造出明亮利落的空间视觉效果。

设计中摒弃了造型与材料的繁复叠加，装修部分以简约的设计理念塑造空间格局，融入温馨、舒适、内敛的设计元素，而装饰方面则运用了加法的设计手法来渲染生活的品位，充分突出了黑白风格的气质——洗练、舒适、自然、明朗。洁白的波浪板，犹如窗外湖水中微微泛起的涟漪，设计师将外景抽象地借入室内，来增加空间的灵动性，正是"鸟鸣山更幽"，与设计的主题——涟漪也相得益彰。

Qin Qin Home Park

亲亲家园

项目面积：140平方米
设计师：周桐
设计公司：杭州周视空间设计机构
主要材料:艺术壁纸、镜子、面板

The designer sets the design concept of this residence as the style that combines the modern household atmosphere with detailed luxury.

To reveal the fashionable aura, the designer chooses the titanium golden color which is integrated with different warm colors as the key tone. The natural patterns of the gray mirrors on the suspended ceiling and sofa background wall makes the harmony of the spaces and shows the lively space atmosphere of the modern rooms. In terms of the choice of the drop light and other furniture, the designer, at the same time, takes the luxury characteristics of details into consideration, and makes the spatial elements have multi-layered impressions. The bedroom takes the modest blue as the main tone, and makes the whole space produce a touch of peacefulness, which just fits the spatial functions of taking a rest. Through design, this project reveals its individuality, and creates a modern and fashionable residential space.

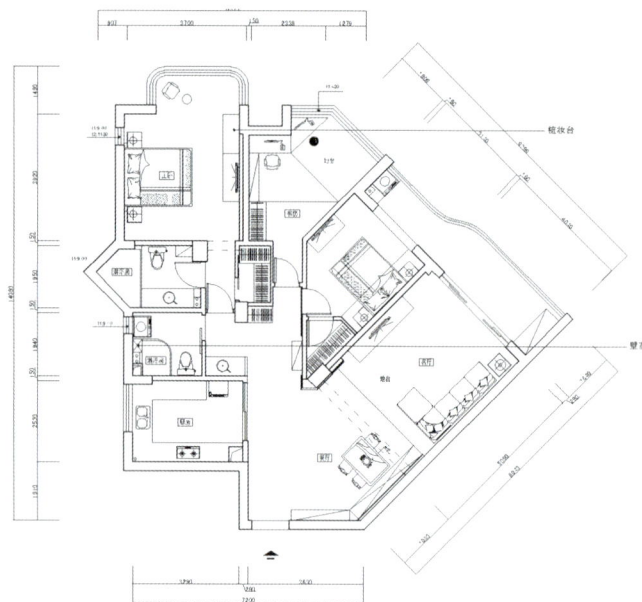

设计师将该居室的设计理念定位为将现代感的居室氛围与细节的奢华相结合的风格。

设计师以各种不同暖色构成的钛金色调为主调，展示时尚气息。吊顶、沙发背景墙的灰镜等处的自然纹样在空间中产生协调，烘托了现代居室富有生气的空间氛围。在吊灯及部分家具的选择上，又同时兼顾了细节的奢华特质，使空间元素具有多重层次的观感。卧室以低调的蓝色色调为主色调，使整个空间氛围产生一丝宁静的感觉，适合休憩的空间功能。通过设计，流露真我的个性，创造现代时尚的家居空间。

The Development of Neoclassicism

新古典演绎

项目地点：台北
项目面积：约132平方米
设计师：张嘉芳
设计公司：真观空间设计有限公司
主要材料：进口大理石、造型玻璃、造型木片、烤漆玻璃

The designer takes different materials on a rich presentation, and makes life have more space for imagination. Through the mutual collocation of natural lights and indirect lights, the designer releases the maximum of the space in a positive and effective way, and at the same time, endows much interest with the space.

The open flat connects large sizes of floor-to-ceiling window, and extends a public space that is between the interior and exterior spaces, so that the warmth of sunlight can embrace the whole apartment. The penetrating tension of lights becomes the main impression of the dining area. With the changes of lights and shadows of the wall, the designer chooses the exquisite totems that are full of classical languages to form the extending themes of the elevation, so that it reaches the extension of the spatial layer. Inside the rooms, the designer selects sober and natural colors as the key tone. The main wall of the living room uses black marble as the background, and indirect illumination to segment the elevation. Matched with symmetrical languages of the wood ornaments, the whole space is embraced with a quiet and balanced quality.

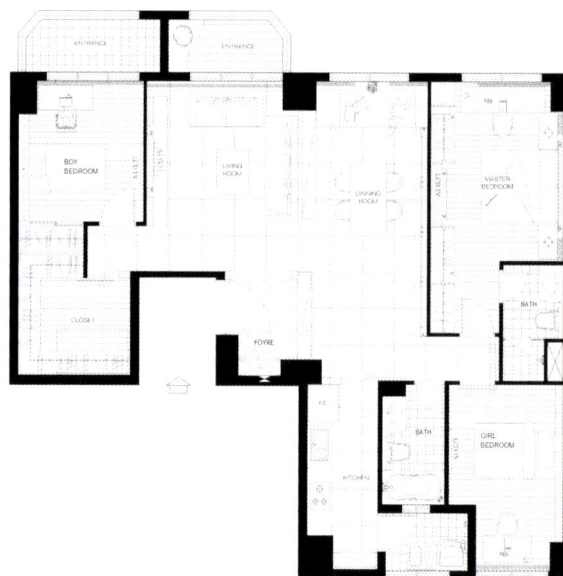

设计师将不同的材质丰富呈现，让生活有更多想象的空间。通过自然光源与间接光源的互相搭配，积极而有效地释放出空间最大值，同时赋予其相当高的空间旨趣。

开放式平面结合大片落地窗，延展出介于室内外的共生空间，让阳光的温度围拢全室。光的穿透张力成为餐厅区域的主要表情，墙面藉由光影的变化，以精美的、颇具古典语汇的图腾形成立面的延续主题，达到空间层次的延伸。室内素材选用沉静、自然的基调。客厅主墙运用黑色大理石做背景，以间接光源分割立面，配合木饰片的对称语汇，使空间具有安定的气质。

Tourmaline

碧玺

项目地点：深圳
项目面积：186平方米
设计师：刘建辉
设计公司：深圳市矩阵室内装饰设计有限公司
主要材料：古木纹大理石、天堂鸟大理石、黑檀木、金箔、软包、手绘墙纸

This project takes the Chinese style as the theme, and uses the contrasting way of material languages to make a fully development for the modern luxurious style. The designer, through the refining and subliming of the modeling conception, designs and treats the interior space.

Black wood patterned stones are like the Chinese painting of those literators and artists, setting off the white stone columns and the background walls that take the scrolls of books as the image; the dark black ebony represents the sober and long-lasting cultural deposits; the golden edges, as the finishing touch, give much emphasis to the hand painted silk embroidered walls and the soft packs with typical patterns; the reflection effect of the black mirror material makes the space full of changes. The whole space is in a sharp contrast and the effect is prominent, which is resplendent and refined, dignified and without losing the human spirit.

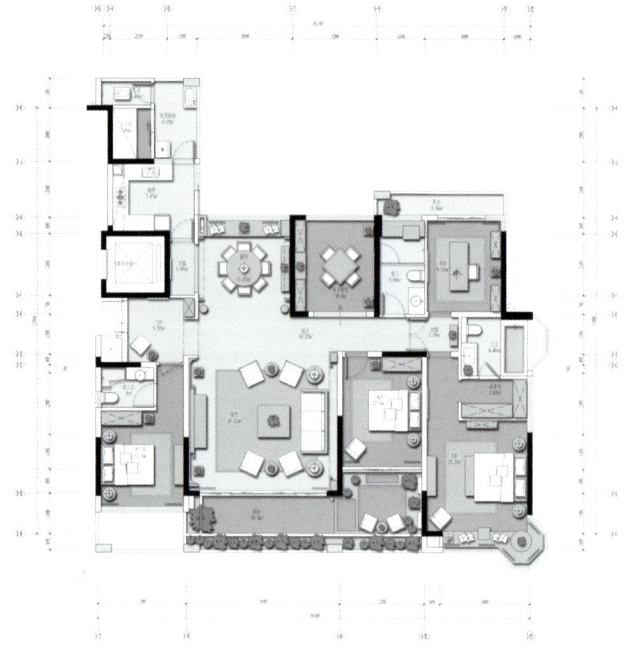

该项目以中式风格为主题，采用了材质语汇的对比手法，将现代的奢华风格演绎得淋漓尽致。设计师更是通过造型意象的提炼和升华来对室内空间进行设计处理。

黑色的木纹石犹如文人笔下的水墨画映衬出以书卷为意象的白色石材柱及背景墙面；深色的黑檀木代表了深沉悠长的文化底蕴；金色的镶边作为点睛之笔，突出表现手绘的绢绣墙面及有代表性纹样的软包；黑色镜面材质的反射效果使空间富于变化。整个空间对比强烈，效果突出，华丽而不失儒雅，厚重而不乏人文精神。

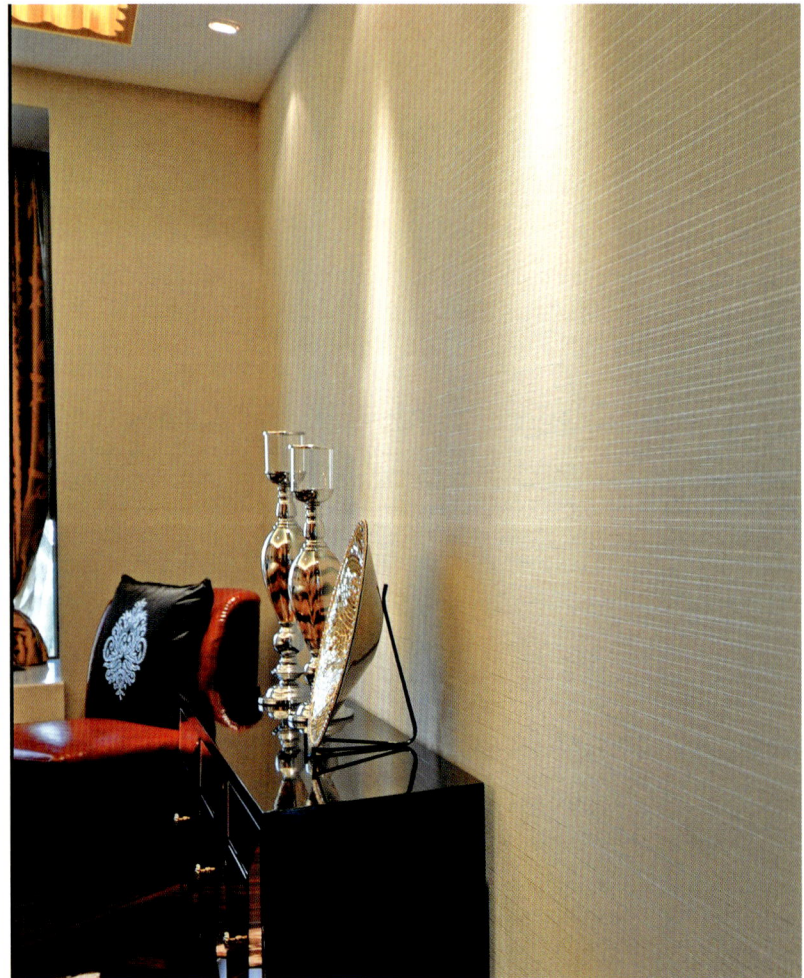

Shiny Valley Show Flat Unit C1

首邑溪谷样板间C1户型

项目地点：北京
项目面积：123平方米
设计师：高敬
设计公司：北京合成行装饰设计工程有限公司

The designer focuses on the interpretation of low-key and gorgeous style in the space, and creates a luxurious atmosphere of space through the selection of materials and carving of details. The interior space is designed with a lot of metallic furniture, and the walls in living room and dining room are decorated with mirrors, echoing with the metallic furniture. The upper and lower seats of six-seat chairs are designed with special forms, forming separation from the other four leather seats, endowing the space with rich changes. The study is designed with deep-color wood floor, coupled with square grid ceiling, making the space tend to be peaceful and cool. The marble mosaic floor defines and guides the spaces through different way of connecting.

本案设计师注重对居室低调而华丽的风格的诠释，通过对材质的选择、细节的雕琢，营造出豪宅一般的居室氛围。室内大量采用了金属质感的家具，在客厅、餐厅等处墙面使用了镜面进行装饰，与金属质感的家具形成了良好的呼应。六人座的餐椅，在上位和下位采用了特殊的形式，与另外四个皮质座椅形成区分，使空间表情更富变化。书房采用了深色实木地板，搭配了井字格天花，使空间氛围趋于平和冷静。地面的大理石铺地通过不同的拼接方式对空间进行界定和引导。

Hong's Residence of Taoyuan

桃园洪邸

项目地点：台湾桃园
项目面积：160平方米
设计师：黄士华、孟羿彣
设计公司：隐巷设计顾问有限公司
参与设计：藏弄设计团队
主要材料：梧桐实木、胡桃实木、烤漆玻璃、莱姆石、南非浅黑烧面、
灰镜、钢琴烤漆、海岛行实木地坪

The natural conditions around the project site are very good, and the three sides of the building feature natural ventilation and lighting. The design uses large French windows to introduce light into the space, making ventilation more flowing. The designers discard the redundant decorative materials and integrate the trivial functions into a whole space, refining the spatial scale and lines and showing the true beauty of the space. The family life focuses on the open public areas, weakening the definition of space attributes and making the moving lines simple and smooth. The interior is decorated with warm touch feeling of natural raw wood and stone material, coupled with effective lighting, leaving the strong characteristics of space image for the owner to play. The design is just like a touch of beautiful scenery, showing the imagination of life.

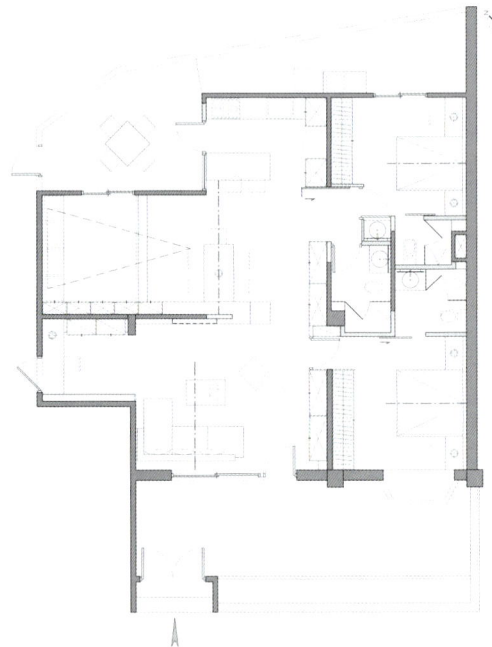

该项目场地的自然条件十分理想，建筑的三面皆有自然通风与采光。设计利用大片落地窗将光线引入室内，使通风更流畅。设计中舍弃了多余的装饰材料，将琐碎的机能整合为一个整体空间，精炼空间尺度与线条，显现空间真实的美感。将家庭生活重心放在开放的公共区域，弱化了界定空间的属性，使动线精简而流畅。室内采用天然原始实木及石材温润的触感，搭配灯光效果，将强烈的空间形象特质留给业主发挥。设计如同一抹优美的景色，映衬着人们对于生活的想象。

Jianhu New Century Garden Show Flat

建湖新世纪花园样板房

项目地点：江苏建湖
项目面积：106平方米
设计师：彭政
设计公司：香港彭政设计师（上海）有限公司
主要材料：白色玻化砖、黑檀木、镜子、墙纸、马赛克、乳胶漆等

The design of the project is classical style, bringing sober and steadfast feeling at home. In the dining room, the wall lamps and droplight light up the dining area commonly, and the glass partition expands the space, creating a sense of permeability in the space. In the living room, the colors of sofa and cushions are unified and coordinated, and the color and style of tea table echoes with the ones of dining chairs .The pictures on the wall and potted landscape on the cabinet endow the living room with some lively atmosphere. In the study, the generous deep color droplight echoes with the deep color floor with natural wood texture, creating a quiet atmosphere in the space. In the master bedroom, two wall lamps are placed on both sides of the main wall, consistent with the symmetry of classical taste. In the wash room, the marble with uniform texture is full of cultural deposits, and the main wall with golden mosaic patterns is gorgeous and elegant.

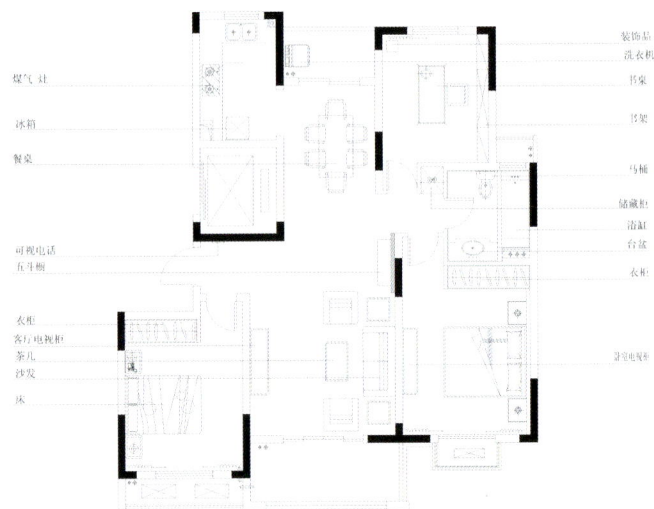

本案设计采用了古典风格，给人带来稳重踏实的居家感受。餐厅的壁灯与吊灯共同照亮了用餐区。玻璃隔断拓展了空间，产生空间的通透感。客厅的沙发和靠枕的颜色统一协调，茶几和餐厅的餐椅的颜色与风格相呼应。墙上的画和柜上的盆景为客厅增添了几分活泼气息。书房大气的深色吊灯与深色天然木纹地板相呼应，营造了沉静的空间氛围。主卧的两盏壁灯分布于主墙面两边，符合古典审美的对称性。卫生间大理石统一的纹理文化底蕴十足，主墙金色拼花马赛克华丽而典雅。

Flower Coast Show Flat

花语岸样板房

项目地点：深圳
项目面积：130平方米
设计师：董泽定
设计公司：深圳翰格环境艺术设计有限公司
主要材料：爵士白大理石、莎安娜大理石、墙纸、白色聚酯漆

The project is simple European style, and the interior is decorated with streamlined European lines with unified style, but not complicated or verbose. The interior is mainly white further highlighting the simple character of the space. The designer highlights the important spaces through adding accessories, and each functional space is led by unified-style droplights with different forms, showing the European style.

The biggest change of plane is to change the original small three-bedroom space into a two-bedroom space, with a master bedroom sufficient to compare with the luxury, giving the owner a totally new feeling. In addition, the open kitchen widens your view, making the originally narrow dining room broaden.

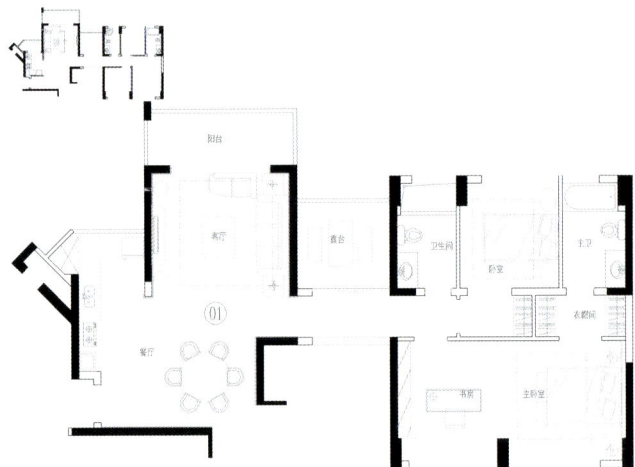

本案采用了简约欧式风格，室内使用精简处理的欧式线脚，风格统一，并不繁复累赘。室内主要采用了白色调，进一步突显了居室简约的性格。在主要空间中，设计师通过增加配饰突出特色，每个功能空间都用形式不同但风格统一的吊灯进行统领，彰显了欧式格调。

平面上最大的改动是将原来的三室格局改为两室，拥有了一个足以和豪宅媲美的主卧，给居者一个全新的感觉。另外，开放式厨房也将视野打开，使得原本局促的餐厅变得开阔。

Conceptual and Environmental Life

意·境生活

项目地点：南京
项目面积：139平方米
设计师：李光政
设计公司：DOLONG董龙设计
主要材料：木纹石、硅藻泥、玻纤布、进口墙纸、夹绢玻璃、马赛克等

A perfect space is like a photo or a poem, which gives off a pleasant beauty and creates an enchanting beauty of artistic conception that makes one become intoxicated.

The design of this project is closely enclosed around this "beauty of artistic conception", vividly describing every corner of the family. The harmonious and integrated neutral colors make the space transparent and bright. The decoration of the foyer is just to the point, and enclosed by green plants, where the real in the unreal and the unreal in the real, which is quite intriguing. In the lamplight, the quality of light and texture of the materials, the massiveness of the furniture and the elaborate decorations of small ornaments make the whole space appeal so natural and rich. In such an atmosphere, the space slowly sends out its individualized charm and produces an upgrade of the artistic conception.

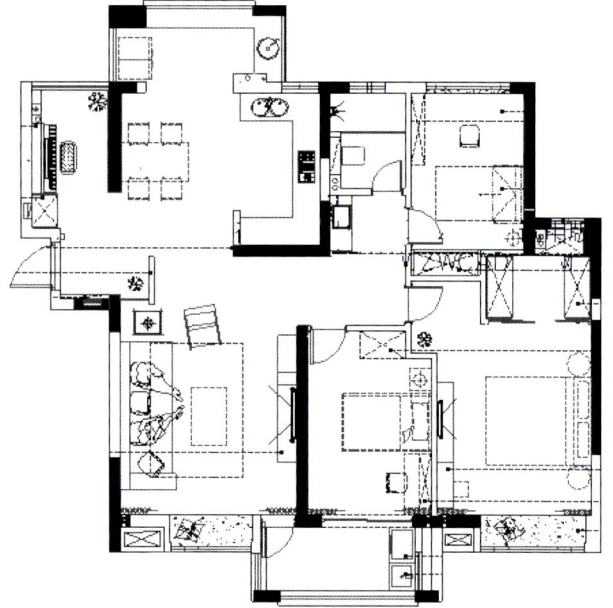

一个完美的空间就如同一幅画，一首诗，散发着赏心悦目的美，营造出的是一种让人沉醉的意境美。

本案的设计紧紧围绕这种"意境美"，生动地描绘了家的每一个角落，和谐统一的中性色调使空间通透明亮。玄关装饰恰到好处，绿色环绕，虚中有实，实中有虚，耐人寻味。灯光下，材质的光感和肌理、家具的体量感和小饰物的精心点缀使整个空间显得如此自然和充实。在这样的氛围中，空间慢慢散发着自己独特的魅力，产生了一种意境的上升。

Ocean Wide International

泛海国际

项目地点：北京
项目面积：360平方米
设计师：顾程
设计公司：尚层装饰（北京）有限公司
主要材料：壁纸、大理石地砖、地毯

The project is neo-classical style design, where the indoor colors are strong and gorgeous, and the decorative style is luxurious and elegant, showing the owner's elegant taste and reflecting the rich cultural heritage as mansion.

In the design, the designer abandons the complex arrangement on the wall in the traditional style, but chooses to reserve some elements of traditional style in the design of curtain and drapery, highlighting the details of decorations. Through the match of colors and materials, the elegant feature of neo-classical style is fully shown out. In addition, the simple and clean lines and selection of fine materials make the whole space more generous.

本案的设计采用了新古典主义的风格形式，室内的色彩浓重而艳丽，装饰风格奢华而典雅，体现了主人的高雅品位，透露出豪宅般深厚的文化底蕴。

设计师在设计中摒弃了传统风格中繁复的墙面布置，而选择在窗帘和帷幔的处理上将一些传统风格的元素稍作保留，让物品的细节更为突出。通过色彩与材质的搭配，将新古典主义典雅的特征展露无遗。另外，简洁的线条和精细材质的选用，也让整个空间氛围更显大气。

Guizhou Bailing Fashionable World, NO.1 Fine Decoration Sample Apartment

贵州百灵时尚天地精装公寓样板房一

项目地点：贵州
项目面积：130.79平方米
设计师：曾涛
设计公司：贵州峰上室内外设计工程有限公司
主要材料：马可波罗砖、罗马利奥砖、墙纸、西奈珍珠石材、石英石、科勒洁具

The project is mainly the modern and simple European style of pragmatism, and it is designed with the living and working lifestyle, focusing on meeting the pursuit of efficiency work and leisure lifestyle for these people.

The interior public areas are decorated with glass as partition, meeting the function of separating and enlarging the space. The different paving materials on the ground naturally separate the areas with different functions, where the public space is paved with floor tiles, and the private space is paved with light-colored wood floor. In the working area, the use of imitated crocodile skin decorative material and the display of European-style furniture enhance the sense of quality in the living space. The bedroom is decorated with gorgeous warm velvet fabric to make the space soft and comfortable.

　　本案以现代简欧的实用主义为主要装饰风格，针对居家办公族的生活习惯进行设计，着重满足于此类人群追求高效工作与舒适休憩二者兼得的生活方式。

　　室内公共区域采用玻璃作为隔断，既满足了分隔的功能，又使空间得以放大。铺地材质的不同自然地区别了各个不同的功能场域：公共性空间采用地砖铺地形式，而私密区域则采用了浅色的木质地板。工作区域仿鳄鱼皮饰材的运用、欧式家具的摆放增加了居室的品质感。卧室运用华丽的暖调丝绒面料，使个人空间柔和又不过于居家。

Issey Miyake

A scent by Issey Miyake

The famous Japanese fashion clothing design masters three home life in 1992, the water creates life style, simple, clean water lily and the integration of the spring flowers in spring, and inject east forest fresh water of life, made of pure emptiness and the zen.

Kai Di City

凯迪城

设计师：成春雷
设计公司：装饰新概念(本色空间设计)
摄影师：monica

Traditionally, household decoration design is used to fixing part of furniture on walls, but this restricts the space for later change, and was more easily to be out of date. As a consequence, an increasing number of consumers choose the simplified and lively design technique that highlights focuses. The largest feature of such designs is to leave sufficient space for imagination for consumers, and leave enough flexible space for the family, so that it can change with the trends accordingly as will. As a result, in the process of designing the household decoration, designers often apply straight lines and simple modeling as the leading design concept, and the key tones tend to choose everlasting popular colors, such as black, white and gray, so that the household will take on a rather neutral plain.

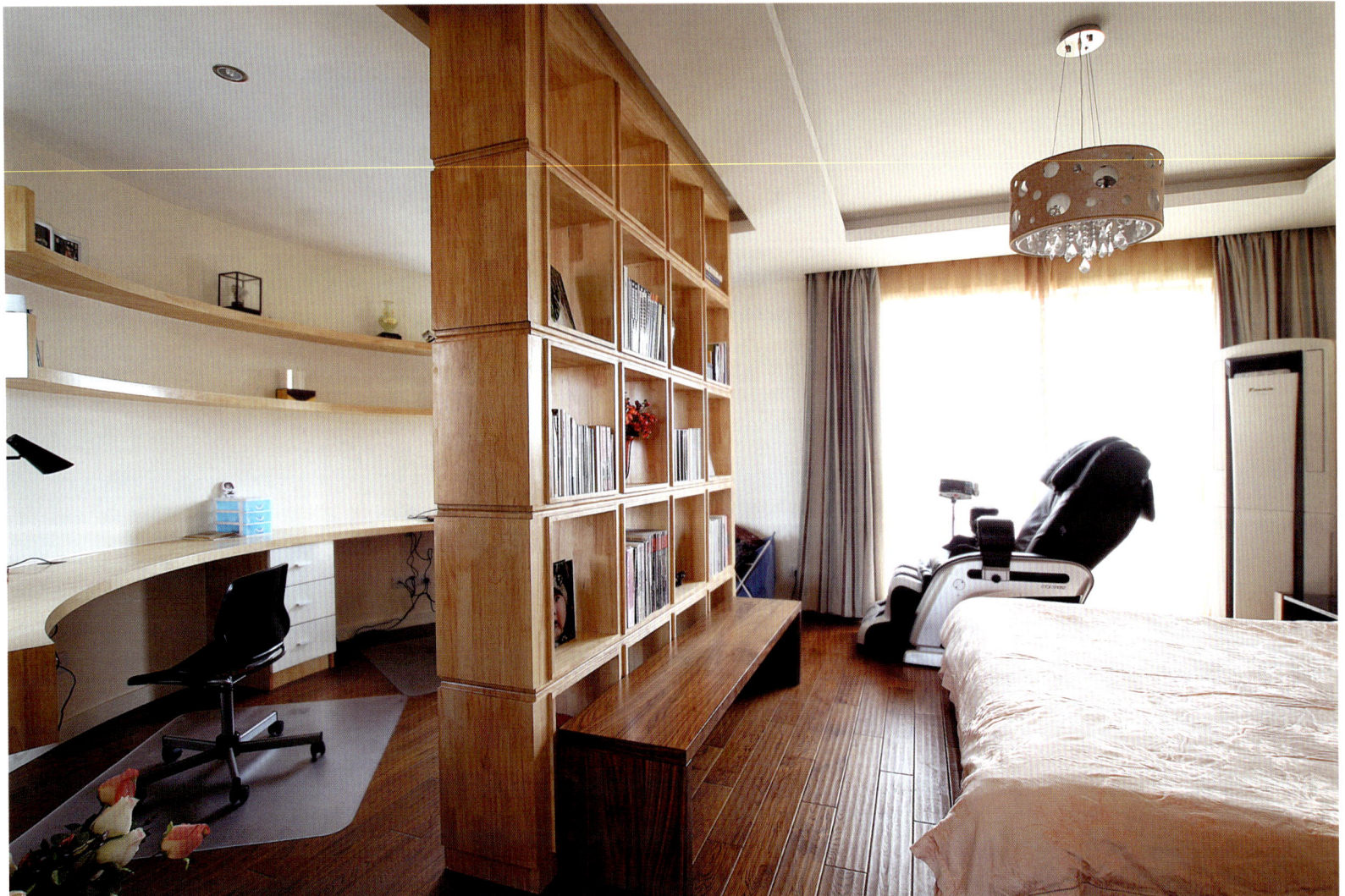

以往的家装设计，一般把部分家具固定在墙上，这样大大制约了房屋后期变化的空间，还容易过时，所以更多的消费者选择简洁轻快、突出重点的设计手法。这种设计的最大特点是留给消费者足够的想象空间，留给家足够的弹性空间，使之可以根据潮流而任意改变。所以在家装设计过程中，设计师往往采取直线条、简单造型为主导设计思路，主色调往往取用黑、白、灰这样的永恒流行色，使室内呈现一种偏中性的质朴感。

Lin's Manor of Dazhi

大直林公馆

项目地点：中国台湾
项目面积：70平方米
设计师：周建志
设计公司：春雨时尚空间设计
主要材料：银狐大理石、镜面、壁纸

In the living room, the silver fox marble is spliced to show modern lines, and the embedded iron accessories fix the TV position, with the use of mirror to lighten the sense of weight of the stone. The sofa background wall in the living room is specially designed with cabinets, where the upside is covered with curtain for dust prevention, and the middle and downside are decorated with double acting doors to separate the storage space, so you can feel the simple and neat way of design. The dining room is decorated with large areas of blue paint glass. In order to weaken the sense of pressure on the roof beams in the original pattern, the designer introduces undulate ceiling form, coupled with embedded lights and indirect lighting to meet the needs of lighting in this area. The master bedroom is designed with soft shape wall, and the headboard is decorated with concave-convex undulate arcs, coupled with double acting doors to feature the TV wall skillfully.

客厅的银狐大理石墙面拼接出现代线条，嵌入式的铁件定位出电视机的位置，镜面的运用使石材的重量感得以减轻。客厅沙发背景墙采用了特殊的柜体设计，上半段用帘子遮掩防尘，中下段用推拉门划分了存储空间，设计手法简洁利落。餐厅空间采用了大面积的蓝色烤漆玻璃。为了削弱原格局中的屋顶梁的压迫感，设计引入了起伏的天花形式，并藉由嵌灯及间接光源满足此区域的照明需求。设计师在主卧空间设计了柔美的造型主墙，在床头做出了凹凸面圆弧起伏，更利用推拉门巧妙地划分了电视墙。

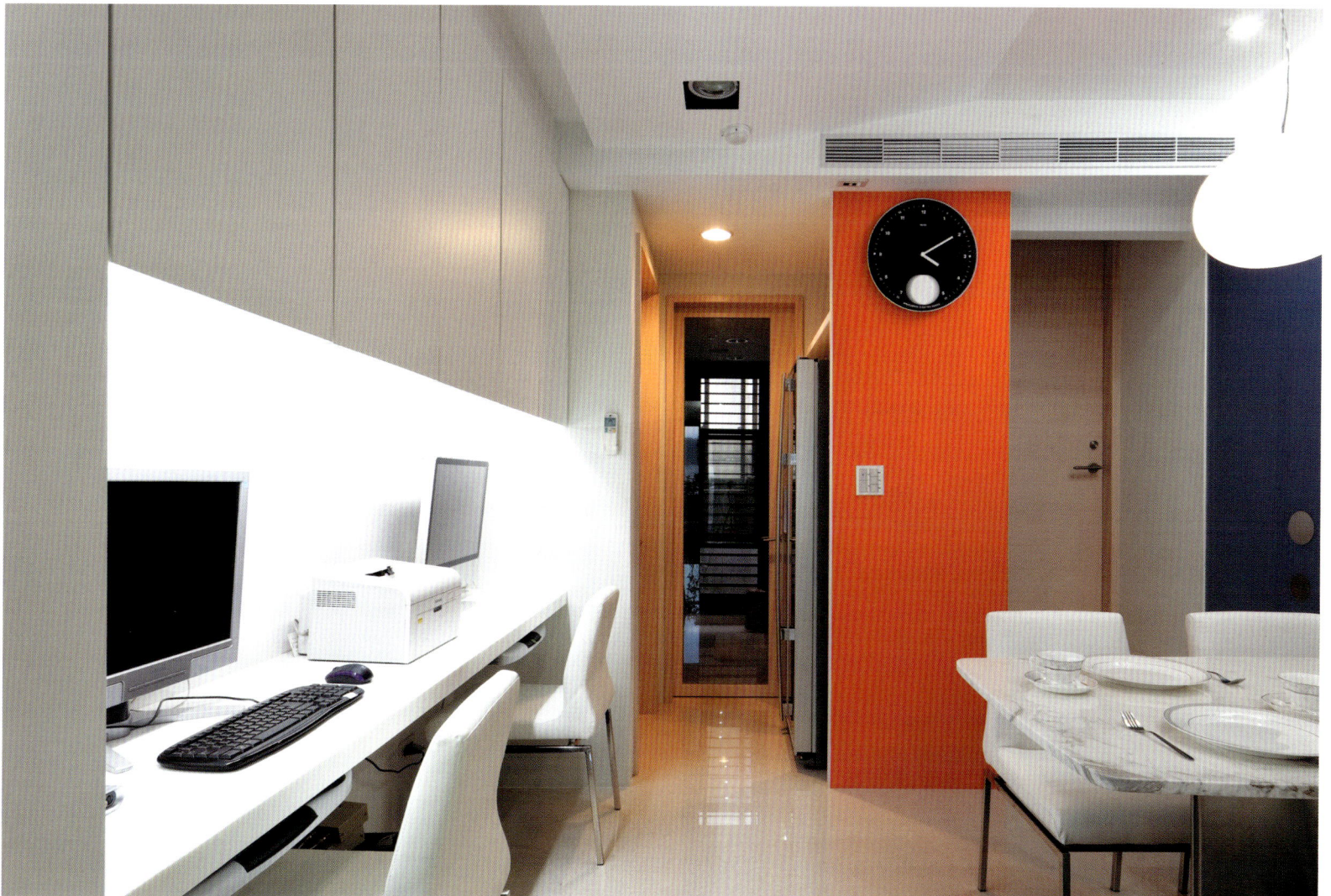

南海南国桃园汇江假日花园A型
设计师：王启贤
设计公司：王启贤设计事务所（香港）

杭州金都高尔夫艺墅
设计师：刘伟婷
设计公司：刘伟婷设计师有限公司

天地凤凰城
设计师：张赫
设计公司：尚层装饰（北京）有限公司

天华美地样板房1008
设计师：彭东生
设计公司：汕头天顺祥设计工作室
摄影师：邱小雄

天荷A10实品屋
设计师：杨焕生
参与设计师：王莉莉、王慧静
设计公司：杨焕生建筑室内设计事务所
摄影师：刘俊杰

旧山顶道
设计师：邓子豪、叶绍雄
设计公司：天豪设计有限公司

君临天华
设计师：陈立风
设计公司：简风设计工作室

木栅黄宅
设计师：杨焕生
参与设计师：郭士豪、王慧静、王莉莉
设计公司：杨焕生建筑室内设计事务所
摄影师：刘俊杰

板桥文邑
设计师：马健凯
设计公司：界阳&大司室内设计

景美蔡公馆
设计师：虞国纶
设计公司：格纶设计

越野豪情
设计师：黄志达
设计公司：黄志达设计师有限公司

保利花园3#C-2米兰印象
设计师：王赟、王小锋
设计公司：广州尚逸装饰设计有限公司

杭州中北花园二期B户型样板间
设计师：徐少娴
设计公司：Gotomaikan International Limited

发光的叶子
设计师：张嘉芳
设计公司：真观空间设计有限公司

细腻空间 暗香盈袖
设计师：吴毅鹏、周颖
设计公司：福州臻美空间设计事务所
摄影师：施凯

台北詹宅
设计师：孙铭逸
设计公司：大宣设计工程公司
摄影师：小雄梁彦

万泰春天
设计师：彭东生
设计公司：汕头市天顺祥设计有限公司

绿光L宅
设计师：唐忠汉
设计公司：近境制作设计有限公司

珊瑚岛奇遇记
设计师：导火牛
摄影师：导火牛

静止的时光
设计师：俞佳宏
设计公司：尚艺室内设计有限公司
摄影师：岑修贤

文质大都会
设计师：张嘉芳
设计公司：真观空间设计有限公司

马他亚住宅
设计师：Elia Felices interiorismo
摄影师：Rafael Vargas

珊瑚阁
设计师：郑勇威
设计公司：爱家设计有限公司

杭州中北花园二期D户型样板间
设计师：徐少娴
设计公司：Gotomaikan International Limited

竹北张公馆
设计师：张巧慧
设计公司：春雨时尚空间设计

混血混搭纯科技
设计师：江先立
设计公司：佶舍室内设计

蓝白色的呼吸
设计师：张静、廖志强
设计公司：之境室内设计事务所

金辉枫尚——木棉家
设计师：陈温斌
设计公司：玄风设计工作室

贵州百灵时尚天地精装公寓样板房三
设计师：曾涛
设计公司：贵州峰上室内外设计工程有限公司

纯境空间
设计师：陈龙
设计公司：南京火龙空间设计工作室

东方名城
设计师：叶强
设计公司：宽北设计机构

贵州百灵时尚天地精装公寓样板房二
设计师：曾涛
设计公司：贵州峰上室内外设计工程有限公司

波托菲诺住宅
设计师：Javier Sánches、Pola Zagga、Larissa
Kadner、Brenda Ochoa
设计公司：JSª
摄影师：Jair Navarrete、Pim Schalkwijk

"白"色浪漫
设计师：吴巍
设计公司：北京东易日盛装饰股份有限公司

贵州百灵时尚天地精装公寓样板房四
设计师：曾涛
设计公司：贵州峰上室内外设计工程有限公司

人·文
设计师：俞佳宏
设计单位：尚艺室内设计有限公司
摄影师：徐凯威

桃园叶公馆
设计师：黄建华、黄建伟
设计公司：黄巢设计工务店

阪急岛公寓
设计师：Elena Galli Giallini
设计公司：ELENA Design Office

黑·白
设计师：蒋兴达
设计公司：福州吉祥如意装饰工程有限公司

新古典风格
设计师：宋建文
设计公司：上海设计年代

银泰中心
设计师：王奇
设计公司：尚层装饰（北京）有限公司

石木水静宅
设计师：林琮然
设计公司：阔合国际有限公司

枫林丽舍
设计师：王品
设计公司：重庆翰艺室内装饰设计工作室

亮贤居
设计师：郑勇威
设计公司：爱家设计有限公司

福州丹枫白鹭
设计师：施继诚
摄影师：施凯

Dolong 董龙设计
Dolong Design & Donglong Studio
[住宅/商业/办公/室内设计]
乐·境
设计师：董龙
设计公司：DOLONG 董龙设计

福州江南水都丽岛
设计师：卓新谛、于斐
设计公司：福州合诚环境艺术有限公司

宜兴景和人家
设计师：宋春吉
设计公司：常熟吉恩设计事务所
摄影师：文宗博

亲亲家园
设计师：周桐
设计公司：杭州周视空间设计机构

新古典演绎
设计师：张嘉芳
设计公司：真观空间设计有限公司

碧玺
设计师：刘建辉
设计公司：深圳市矩阵室内装饰设计有限公司

首邑溪谷样板间C1户型
设计师：高敬
设计公司：北京合成行装饰设计工程有限公司

桃园洪邸
设计师：黄士华、孟羿彣
设计公司：隐巷设计顾问有限公司

建湖新世纪花园样板房
设计师：彭政
设计公司：香港彭政设计师（上海）有限公司

花语岸样板房
设计师：董泽定
设计公司：深圳翰格环境艺术设计有限公司

Dolong 董龙设计
Dolong Design & Donglong Studio
[住宅/商业/办公/室内设计]
意·境生活
设计师：李光政
设计公司：DOLONG董龙设计

泛海国际
设计师：顾程
设计公司：尚层装饰（北京）有限公司

贵州百灵时尚天地精装公寓样板房一
设计师：曾涛
设计公司：贵州峰上室内外设计工程有限公司

凯迪城
设计师：成春雷
设计公司：装饰新概念(本色空间设计)
摄影师：monica

大直林公馆
设计师：周建志
设计公司：春雨时尚空间设计